WIND TUNNELS AND THEORIC CFD: DESIGN AND USEFUL

It is the first book in a large and special series of books, dedicated to motorsport in general; it will cover aerodynamics, suspension, engines, dynamics, etc. ..... Everything you need to learn how to design a full car.

The aim of this series is also to say that I would like to teach again in a university.

I hope that this series will be a success and that I will be able to transmit all my knowledge and all my experience.

@TimoteoBriet

Wind Tunnels

DEFINITION AND TYPES

Already said by Leonardo Da Vinci:

*"Both the thing moves
against the air as the air against
the thing."*

This is the fundamental principle of a wind tunnel; to simulate reality, we do the opposite of reality:

On the real track the only thing that is moving is the car; in a wind tunnel, the only stationary thing is the car; the ground, the wheels and the air is moving.

To simulate different values of speed and air density, we can use and work with the Reynolds number; This number tells us that if we want to simulate a car at a speed and on a scale 1: 2, we must increase the air speed at twice the speed study; if instead we simulate what happens with water, we do the test at least 1000 times speed (the density of water is 1000 times greater than that of air).

Depending on how fast you need to perform in test we use 3 different test tunnels:

Low speed (less than 400 km/h):

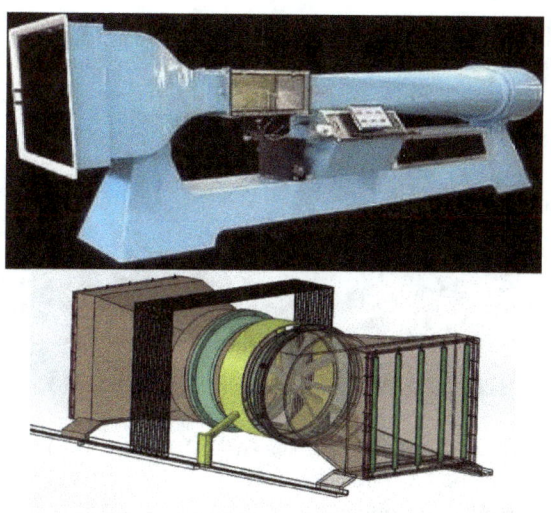

High speed:

Water tunnels: using them, we can test high speeds with cars of small scales:

From the Reynolds number, we increase the pressure inside the tunnel, so that if we increase pressure twice, we can do the test at 1: 2 car with the same speed; It is very difficult to maintain the seal in these tunnels:

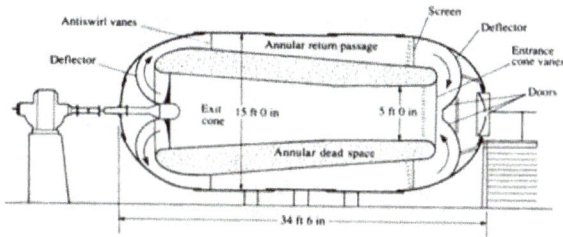

Let's focus on the first type, as all our tests run at low speeds.

Within such tunnels, there are 2 types:

- Open or Eiffel type: They suck air from the atmosphere directly.
- Closed or cyclic; they are essentially Eiffel tunnels but communicating entry and exit.

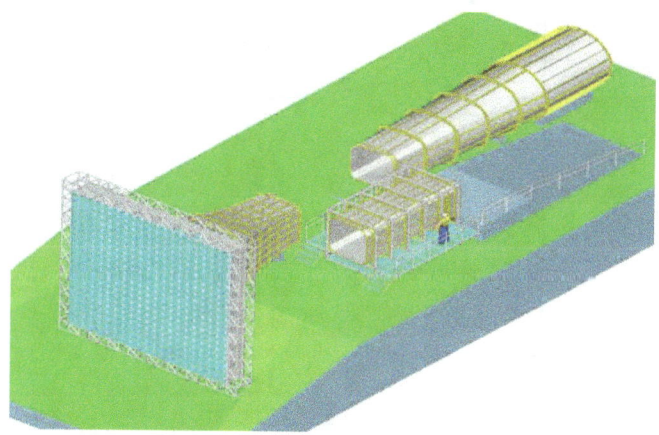

In others cases an Eiffel tunnel is placed in a closed warehouse, making the Eiffel tunnel a closed loop type and dual path also:

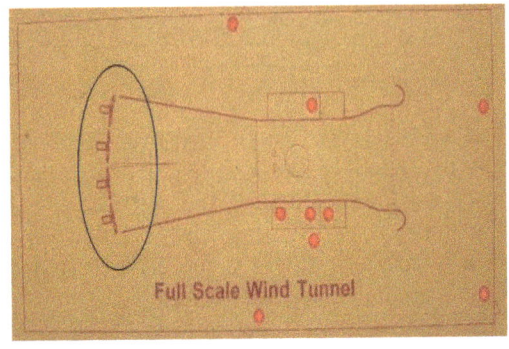

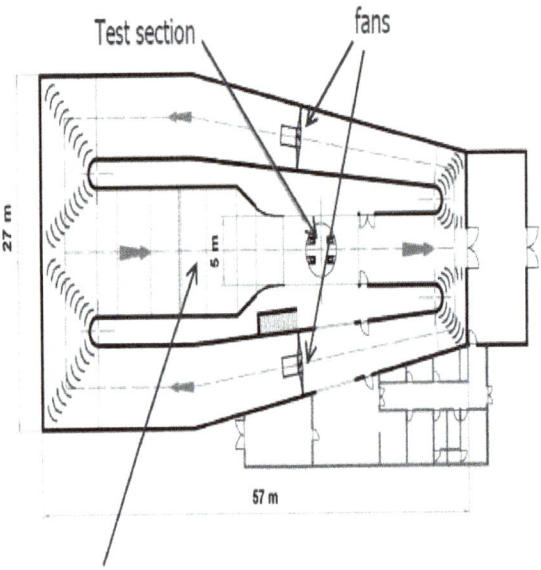

PROS AND CONS

➔ Open or Eiffel:

settling chamber (honeycomb & screens)

test section

flow

fan

contraction

fan   Test section   Settling chamber

airflow

diffuser

Pros:
- Low cost of construction.
- Few usage problems.

Cons:
- The entry determines the quality of flow in the test chamber.
- More power to move.
- Noisy.

➔ Closed cycle:

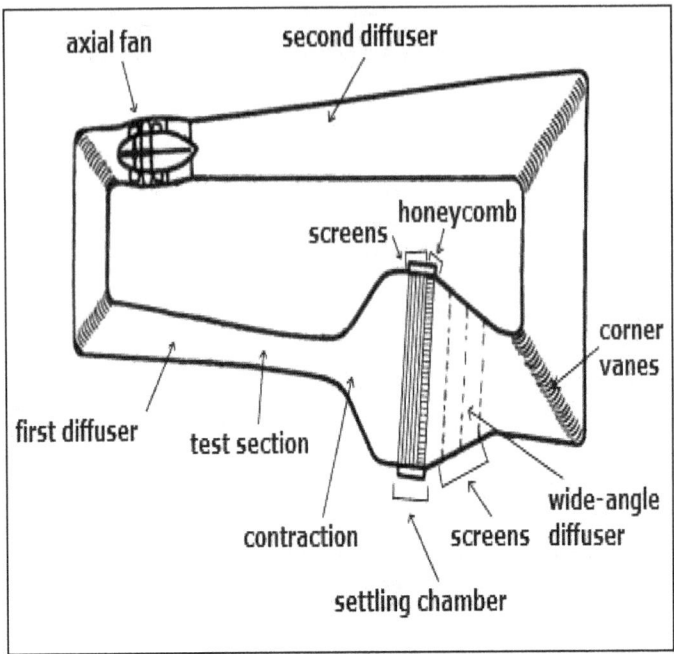

Pros:

- The flow can be controlled.
- Less energy to move.
- Less noise.

Cons:

- Higher budget.
- Air temperature control system is required.

These pros and cons of each type of tunnels are obvious; what really matters are the problems and advantages of the fact of using a wind tunnel, whatever its type.

Letter or work scheme with a "WT":

Scale model WT program
Development flow chart

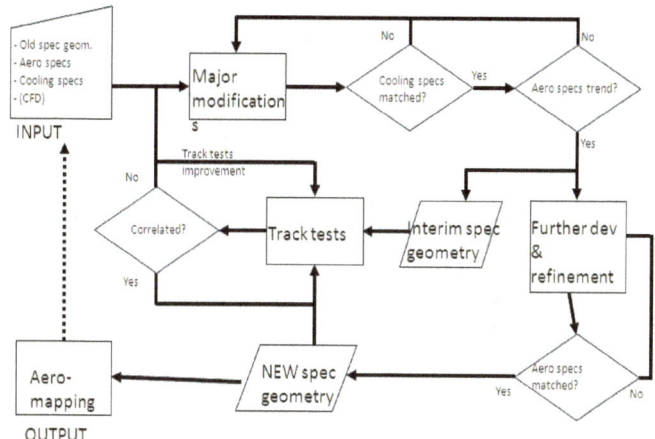

Pros:

- It gets almost the correct prediction of the aerodynamic coefficients, other than those strictly related to the phenomena that depend on Reynolds.
- Repeatable and reliable test experimental conditions (controlled environment tests).
- Huge amount of post-process data quickly available (forces, moments, pressures, mass flow, temperature).
- A large number of runs for each day of the session.

Cons:

- It is difficult to adequately simulate actual conditions (earth moving, blocking test section, etc).
- The volume of the available section for the proper performance of flow visualization techniques is limited and

small.
- The data for the analysis, calibration and maintenance must be taken into account before the test.

→ Car scale 1: 1 real size:

Pros:

- Suitable even to investigate the phenomena that depend on Reynolds.
- Allow adequate flow visualization techniques.

Cons:

- Accurate simulation of rolling land is not common.
- The larger the wind tunnel, the higher the price of wind tunnel time.
- Changes to the car are expensive and costly, in response to evidence of poor design or an improved design. The combination with CFD, is ideal.

Mainly the cost of a wind tunnel, despite being very good for testing, is what limits its use.

→ Reduced scale car:

Pros:

- Fast and relatively cheap to build and modify during test sessions.
- For wind tunnels full-scale sectional area of test can reduce costs as a result.
- Components of high modularity.

Cons:

- In order to maintain reasonable similarity with Reynolds number, the wind speed has to be increased and therefore the stiffness requirements model.
- Not suitable to changes depending on the number of Reynolds.
- The difficulties in simulating the small details of the original form.
- The measuring instrumentation may be limited by the dimensions of the model.

A small scale car represents the most cost-result compromise between data accuracy and fast measurement relationship.

Let us now see compared to tests on track, the pros and cons of this last test:

→ Track test:

Pros:

- It is the actual test.
- The effects of the floor under the car, play back correctly (it is the reality ....).
- Performance comfort (handling) can be simulated.
- You can have the pilot commenting sensations.
- Used to validate data from the wind tunnel.

Cons:

- Difficulty in acquiring data properly; especially in rain.

- To generate aerodynamic data, lots of trials (time and money) is needed.
- It keeps the car "busy".
- The drag is difficult to determine by this type of test.

With all that said, we get an idea of the advantages and disadvantages of each type of test has.
In any case, an optimum combination of:
- CFD.
- Wind tunnel.
- Track.

It is the best and most advisable.

→ Is common, after the fan in closed tunnel for example, increase the duct section in order to:

- Have less friction between air and walls; so less heat....
- Have less thickness boundary layer.
- So: less electricity energy power iiii

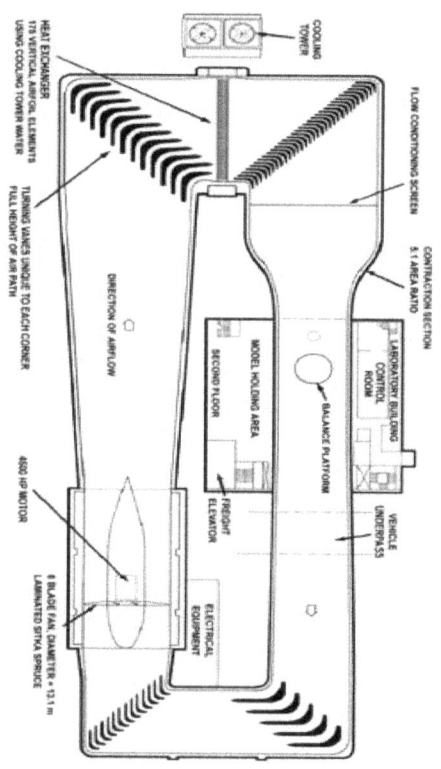

Turning vanes in corners:

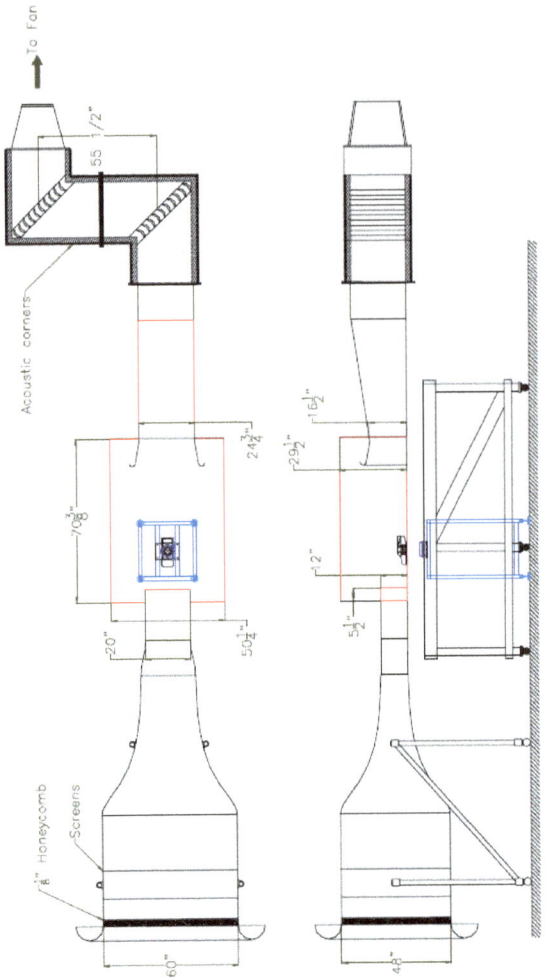

If the friction is big, normally, is necessary to install radiators for reducing or controlling the air temperature; that is very very important:

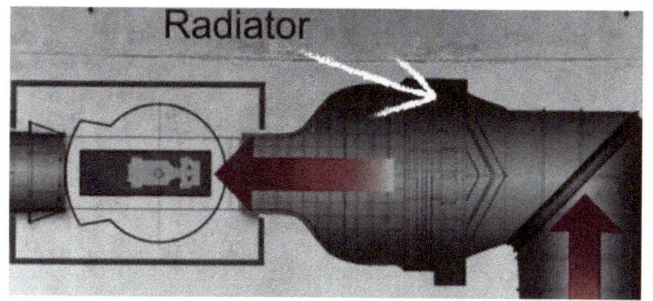

Another think very important to reduce the power wind tunnel, is reduce the energy in corners: these devices, are very importants:

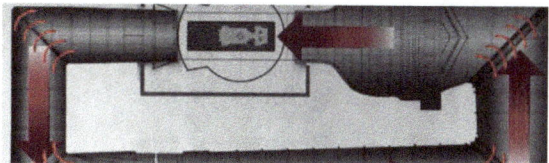

STRUCTURE AND PARTS; DIMENSIONS

A wind tunnel, hereinafter "WT" always has 3 distinct parts:

- Input: is responsible for making the airflow through properly; that is, the flow in the test chamber is laminar.
- Chamber of tests: This is where the car is registered to perform the test.
- Diffuser: in charge of extracting the air and is where the power plant or engines are placed; It is also an area where turbulence should not produce.

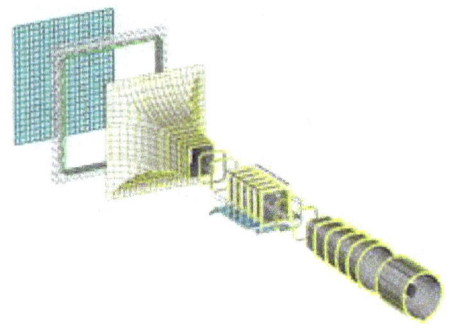

The entry has that special funnel shape which, mathematically you can draw: is an expression of the polynomial function. Its shape is that which causes the flow entering the test chamber to be laminar:

$$L := 1.2 \qquad Hi := 3.5 \qquad H0 := 1.1$$

$$h1(x) := \left[(-10) \cdot \left(\frac{x}{L} \right)^3 + 15 \cdot \left(\frac{x}{L} \right)^4 - 6 \cdot \left(\frac{x}{L} \right)^5 \right] \cdot (Hi - H0) + Hi$$

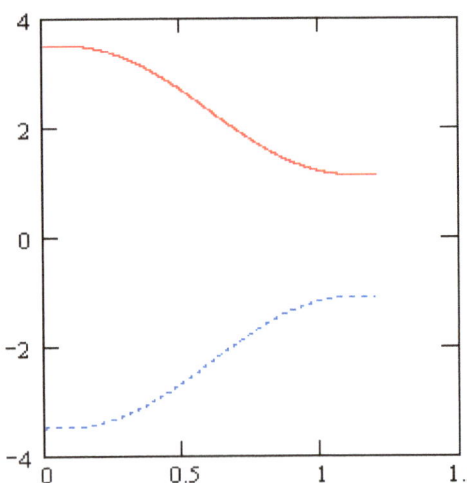

Once we have "predefined" the shape, CFD studies are done to get the full laminarity.

When we have "primary" general shape of the tunnel, begin to study the full "WT" using CFD techniques; remember that what is intended is that in the test chamber, the flow is as laminar as possible.

Normally, when you enter air flow from "outside" over which you have no control, a series of grids are arranged in the front to mitigate and laminar possible great turbulences and even stop the inclusion of birds ... .:

Is possible to help to allow this laminarity, from honey comb panel in inlet test section, and also panels with holes more little's (reduction big turbulences and smalls turbulences):

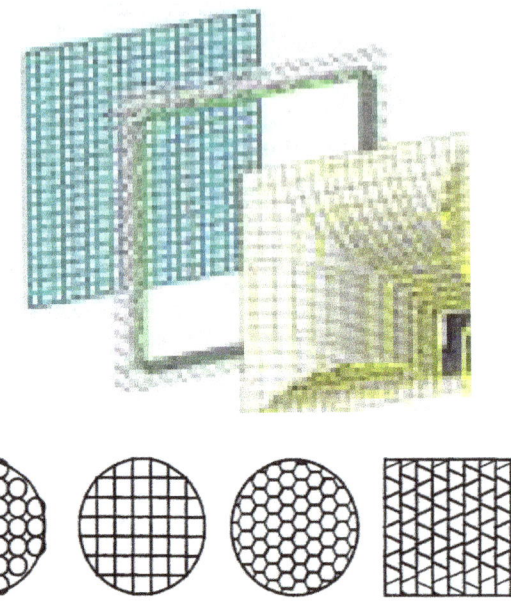

The rate of turbulence in the test section is expressed as a percentage: u, v and w are the speeds in 3 axes, and "U" the full speed vector:

Dans la grande majorité des cas pratiques de véhicules de compétition automobile, le **taux de turbulences moyen** auquel on peut s'attendre lorsque l'on se trouve en condition de course sur un circuit exposé à un vent latéral moyen **est de l'ordre de 0,05% à 0,2 %**, ce pourcentage étant calculé de la manière suivante :

$$\frac{\sqrt{du^2 + dv^2 + dw^2}}{|speed|} * 100$$

u, v and w, are the coordinates of speed; this value is a percentage of turbulence; the value ideal is for dv and dw, zero; we suppose that the flux have only the component "u" or horizontal.
With regard to the test chamber, can be of 2 types:

- The installed in the duct itself.
- Those who are inside an airtight room,

installed as part of the main line. In creating depression, air enters one end of the room impacting on the car and out the other end.

The diffuser is responsible for extracting the air circulating in the "WT"; at the end of the diffuser sucking fans are placed; the diffuser must have walls at an angle which should not exceed 6º. So not peel off the flow of the walls and possible turbulence are reduced.

Regarding the necessary power, a "WT" with a section of 2x2 meters test chamber air circulating at a speed of 200 km/h, it needs approximately 400 kw electric. The relationship between test section room and power, is linear and the relationship between power and speed is cubic:

- If you want a "WT" with a test chamber of 4x3 meters at a speed of 200 km/h, we need a power of 1.2 megawatt.
- If you want a "WT" with a section of 2x2 m test room at a speed of 300 km/h, we will need 1.35 megawatts of power.

It is a simple way to calculate power needed for a "WT": see the size of people in the image below:

The dimensions of a "WT" are determined by different values:

- The scale at which you want to test the car.
- The dimensions of the room where will be installed.
- The performance to be achieved by the "WT" (mainly speed).
- If it is a blown or sucked "WT": in the case of being blown "WT" is much shorter; flow laminators should be placed at the entrance of the test chamber.

By way of curiosity, to say that during a preliminary design "WT" of a university in which we participate, it was wanted to design a "WT" for a speed of 360 km/h, closed-cycle car scale 1: 1; bestial project, really. It was ruled by budget issues, because the system for cooling the air was much more expensive than the tunnel without such a system: the air, after a few minutes, was about 80 ° C warm, which involved the installation of a cooling system.

In reality, the car does not have walls around them; therefore, in a test tunnel, the car and its aerodynamic should not be affected by the presence of walls around it, which significantly distort the data or results determined by the test:

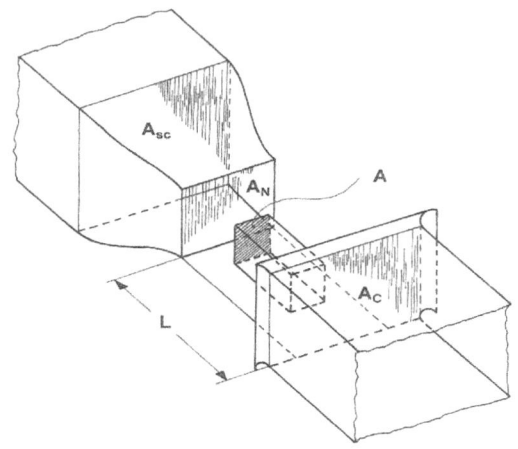

Relative length: $l= L/D_N$
Blockage ratio: $j=A/A_N$
(Reasonable limit for cars: j=0.1)

These dimensions are very important, because for example, if the tunnel section is square, the flow is not the same along full geometry, and can produce a correction values aero, from the named Corrections Blockage (values of turbulence (%) in every zone):

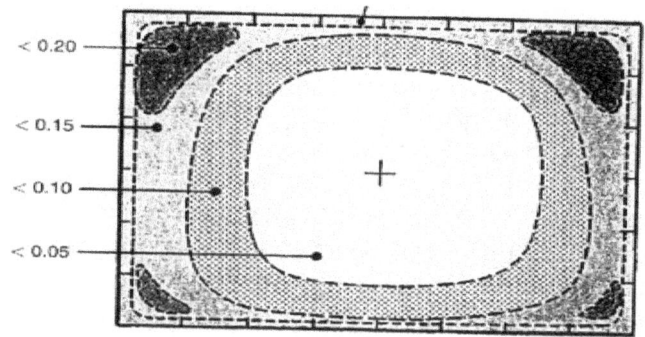

That is, the flow obstruction caused by the existence of a car, must not exceed 10% of the total area of the test chamber; thus, the existence of these walls does not influence on the aerodynamic performance of the car or parameters:

Without walls and with walls (streamlines) – (track and wind tunnel) → streamlines compressed:

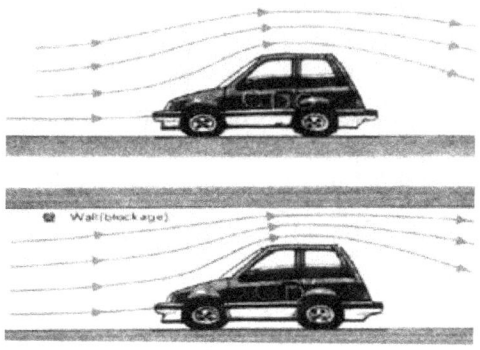

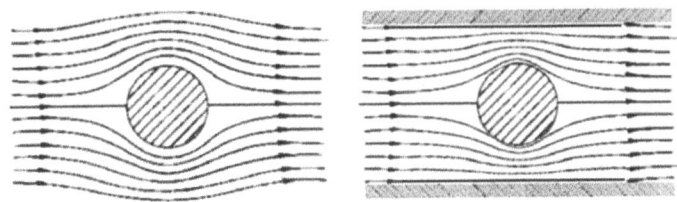

This "modification" of streamlines (and so the pressure field), produce one modification (error) of forces; is necessary so, to define the Blockage Factor correction.

We can test a new roof design; that is (WALL-INTERFERENCE CORRECTION FOR THE SURFACE-PRESSURE HYSTERESIS OF AN AIRFOIL UNDERGOING PLUNGING MOTION: Mohammad Saeedi, Mahmoud Mani and Armin Hamta, Mechanical Engineering Department, University of Manitoba, Winnipeg, MB, Canada, Aerospace Engineering Department, Amirkabir University of Technology, Tehran, Iran):

Slotted roof
(30% porous)

We can test a profile, vibrating in heave (sinusoidal vibration), and e can sse the differents results between rof traditional and this new design:

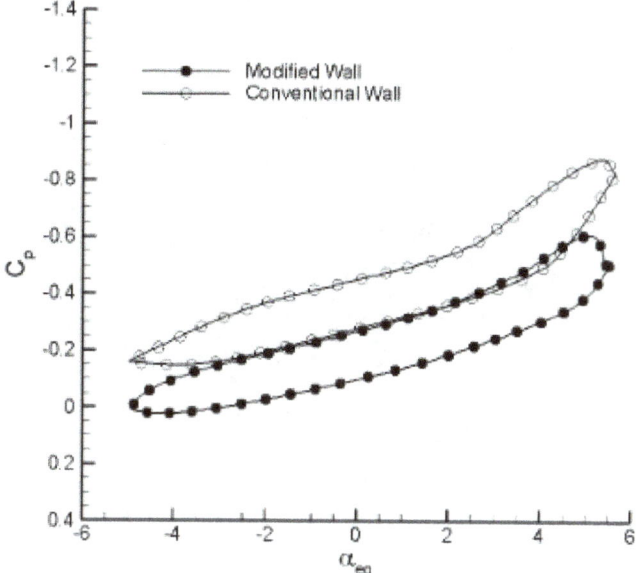

"k" is (wc/(2U)); "w" is velocity angular based on the plunging frequency, "c" is the airfoil mean chord line and "U" is the velocity.
Correction factor blockage:

$$\Delta = \left(C_{p\,\text{modified wall}} \right) - \left(C_{p\,\text{conventional wall}} \right)$$

The blockage in a wind tunnel, is some very important.

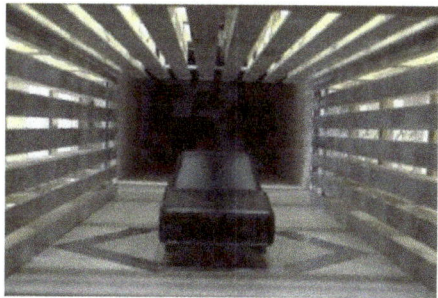

Is possible to use the next correction factors:

"Fwt" can be 0.5, 1.3 or 2.5.

$$C_D = C_D^* \left(1 - Fwt * blockage\right)$$

The best, is to have a test "real" to compare....
There are some systems to slotted the wind tunnel walls:

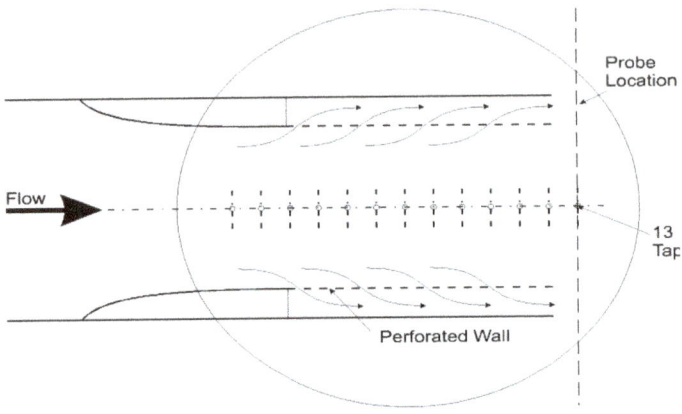

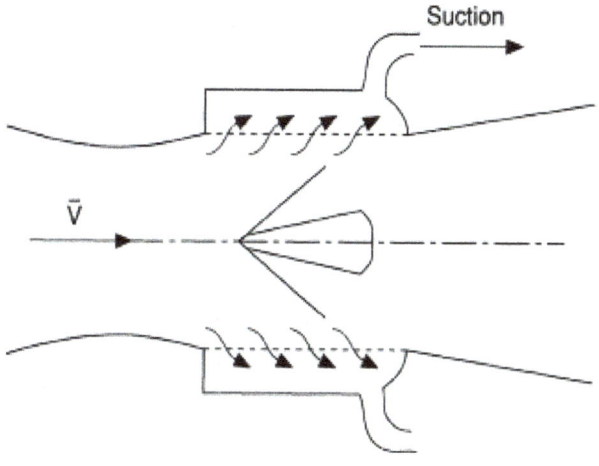

ROLLING FLOOR AND WHEELS; BOUNDARY LAYER CONTROL

For racing cars, a "WT" that does not have rolling floor is useless; With passenger car this inutility is not given; this is because the race cars are very close to the ground with what is basic to simulate ground effect and all its consequences.

In the case of not having the "WT" capacity to rotate the wheels, you can install the following device:

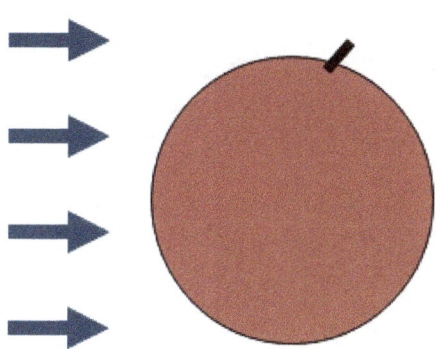

It is a flat plate bent about 45 ° of about 8 cm; this board, "simulates" the existence of rotation of the wheels, without rotating the wheel.... Superb home method, but it works perfectly. The essence is that reproduces downstream turbulence and drag of the wheel.

In the case of the "WT" does not have rolling ground, it is not so easy.

A rolling ground is nothing but a rolling belt through 2 big rollers; to simulate different "circuits and roughness," the tape is changed for another that meets the specifications of the asphalt of the circuit:

If the floor does not rotate, the velocity profile on him would be:

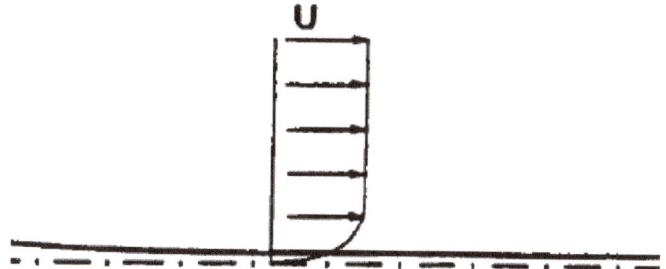

We must try that the profile is (see next image); that is the correct velocity profile:

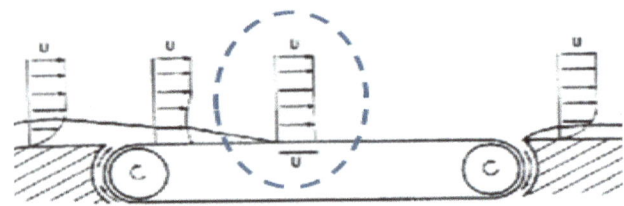

Thus, the velocity profile is perpendicular (no speed variation with respect to distance to the road); this happens on the track, as floor and air are mutually stopped.

But the installation of this moving ground have a some problems (big big problems):

- The ground moving, is heating a lot (refrigeration so).
- The ground is suctioned from up (from car) (suction down).
- The ground have little durability....
- Also is necessary install a suction boundary layer, in order to eliminate this layers, before ground moving:

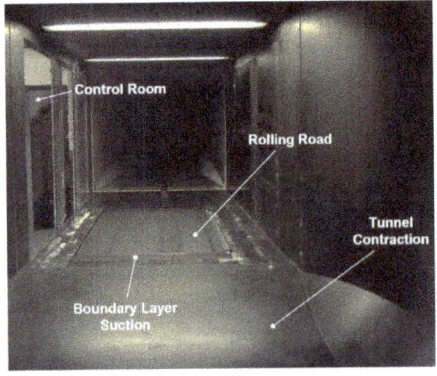

If not exist ground moving, can be see the next (formation air bubble, which affect boundary layer thickness, etc....):

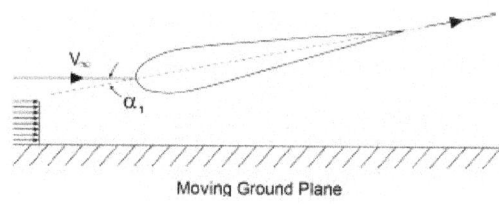

(a) Rolling road on

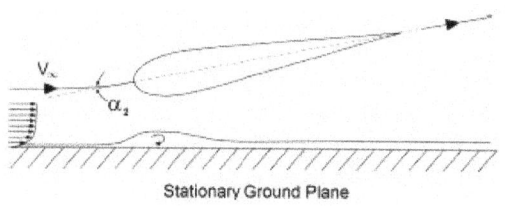

(b) Rolling road off

If there are no rolling ground, we can choose four solutions:

- Raise the car on a platform (with all that that entails of difficulty of "measuring" parameter); the boundary layer starts near the car so when it reaches the thickness of it is "minimal"; It is a solution that mitigates or just hides the error: is necessary that there is the rolling ground.

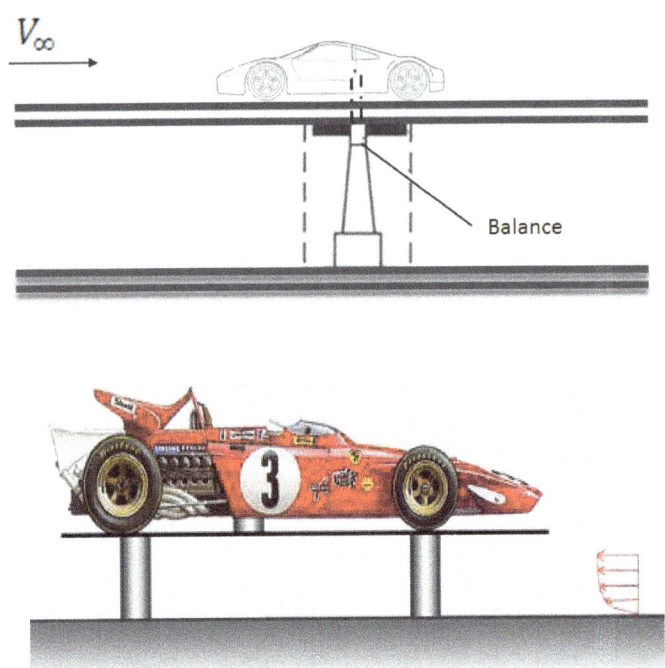

If we cannot realize this complicated system, it is possible to "cut and remove" the boundary layer created in front of the car so that it also generates another near the car but with much less thick. Such is the case of tunnels with a large distance between the air inlet and the car:

- Extract the car down, simulating the "real" ground effect:

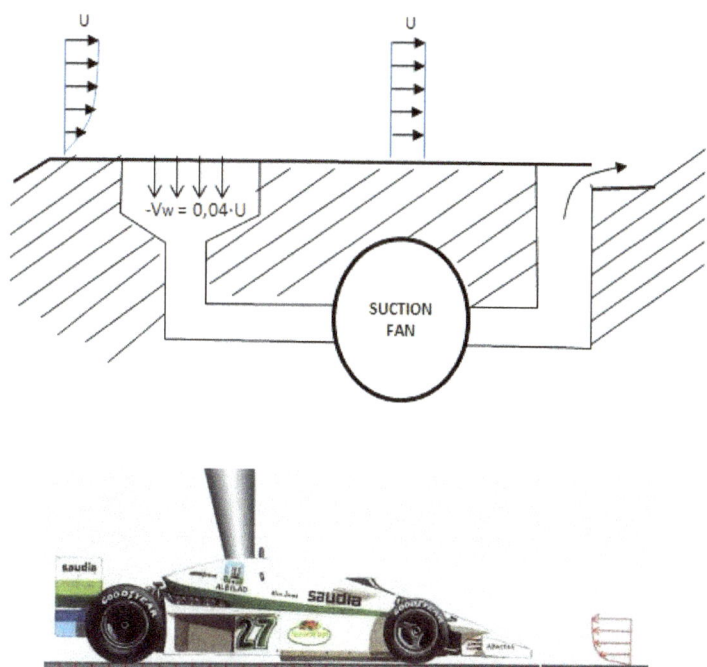

- Inject air at high speed before the floor, and in this way "drag" the velocity profile:

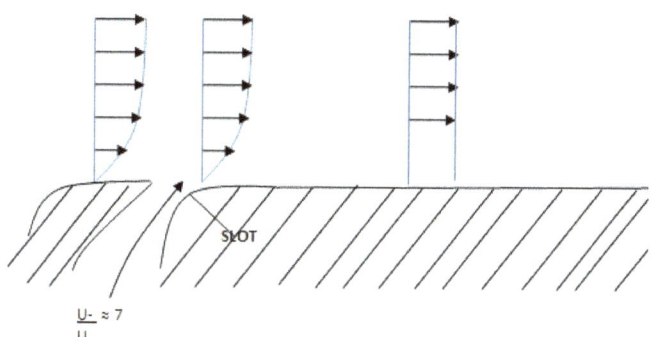

With injection, the value is constant and "1":

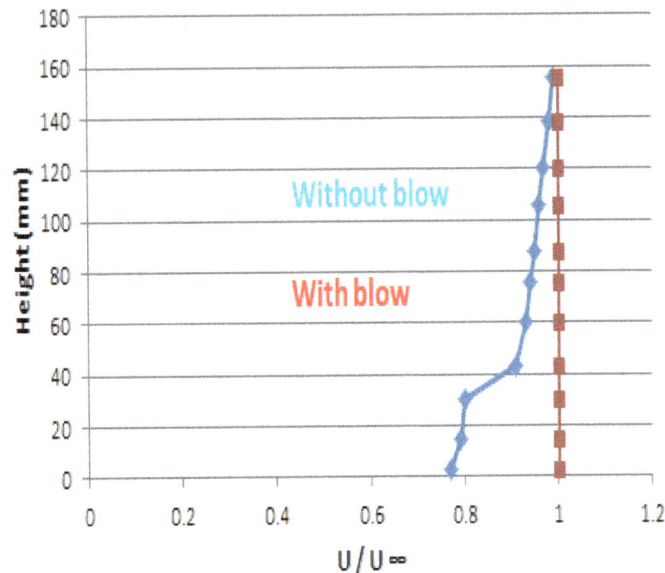

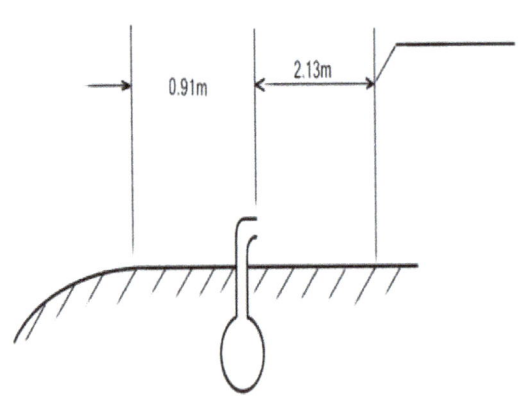

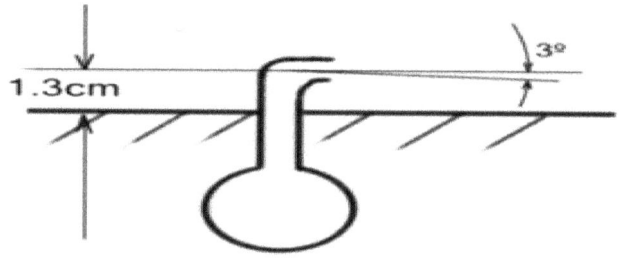

1.3cm

3º

Distance between nozzles: 127 mm.

- Extract the boundary layer at the end of the rolling ground to achieve the same objective (drag the velocity profile):

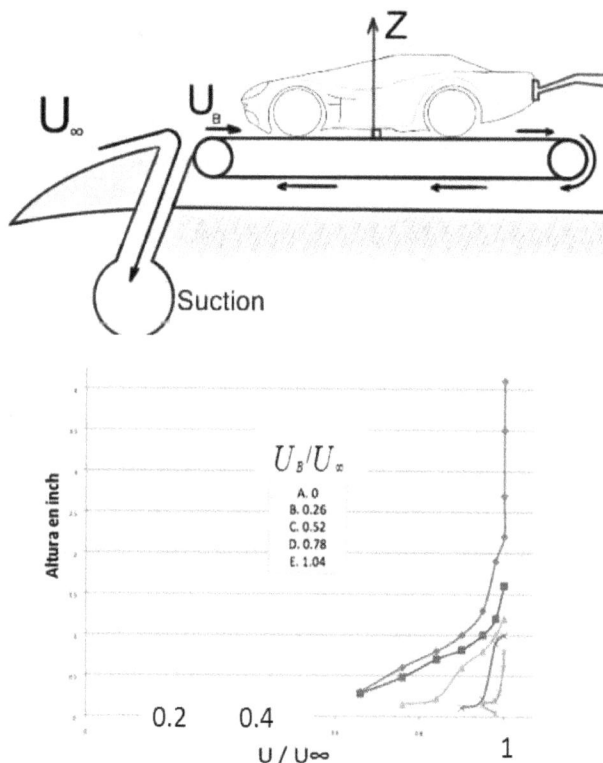

U_∞ U_B Z

Suption

Altura en inch

U_B/U_∞

A. 0
B. 0.26
C. 0.52
D. 0.78
E. 1.04

0.2   0.4

U / U∞   1

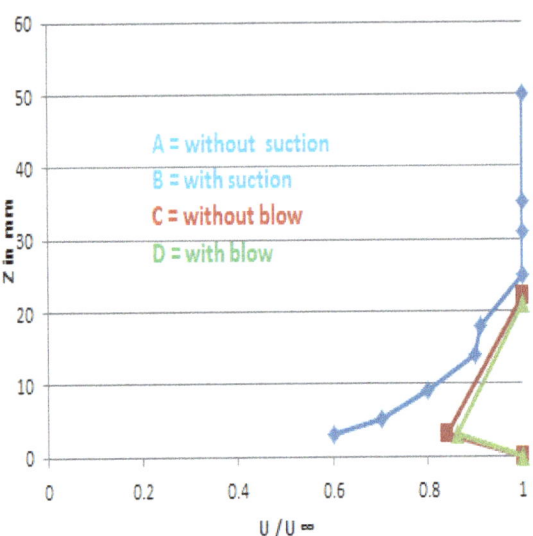

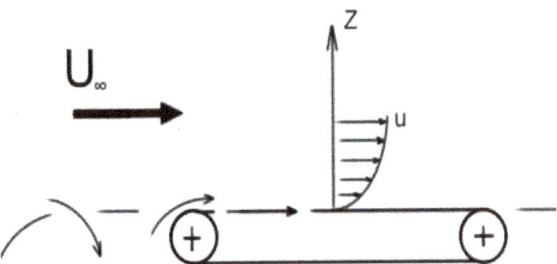

Suction flow:

$$\dot{Q} = v_s \cdot b \cdot s$$

Volumetric flow deficit without the boundary layer
(s = slot width, b = slot length):

$$\dot{V} = U_\infty \cdot b \cdot \delta_{1x=0}$$

Suction parameter:

$$C_{\dot{Q}} = \frac{\dot{Q}}{\dot{V}} = \frac{v_s \cdot s}{U_\infty \cdot \delta_{1x=0}}$$

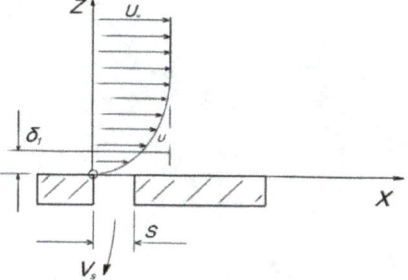

We know the importance of the existence of rolling floor and the rolling wheels when testing a model in a wind tunnel. Here is a small quantification of the coefficient change of pressure over the ground and the diffuser of an F1, with and without rolling floor and rolling wheels:

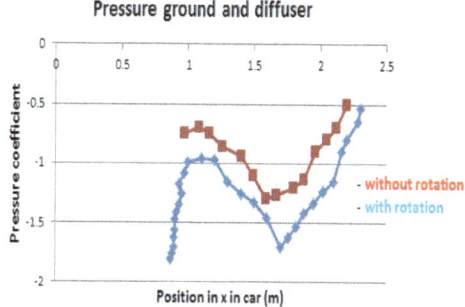

Appreciate the fact **of** rotate both systems involves an increase of downforce quite appreciably.

- Another option in order to simulate the ground effect, is:

OTHER IMPORTANT SYSTEMS

There are countless systems installed on a "WT":

- Changing the ribbon to simulate different roughness circuit.
- Turbulators at the entrance of the test chamber to simulate turbulence or

cars in front of others:

- The platform where the car rests can rotate and swing, to simulate roll and yaw.
- A myriad of other systems.

To see the influence of the yaw of the car in its aerodynamic parameters, consider the following example (the infrastructure to place measurement and movement, is spectacular):

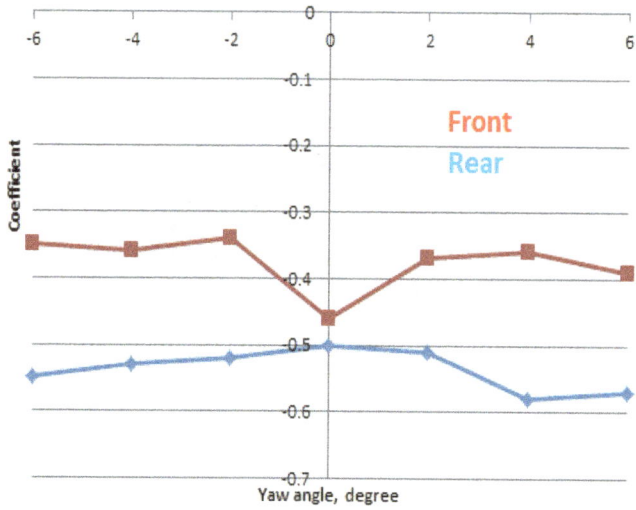

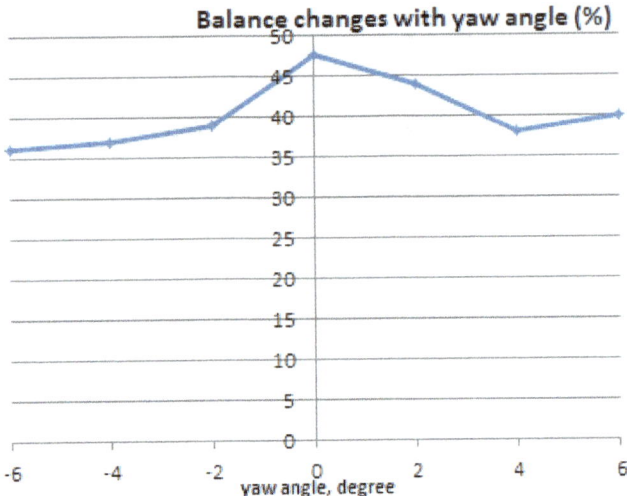

Balance changes with yaw angle (%)

We see that with small changes in yaw, the aerodynamic performances also change.

You can also change the pitch of the car; an important test in terms of risk prevention; take notice that in an accident, one of the attitudes of the car is just pitch + yaw.

Moreover, we can combine yaw variation with variation of pitch and roll:

For that, open test room is ideal; if is not possible, is necessary to install holes in walls test room (blockage effect also):

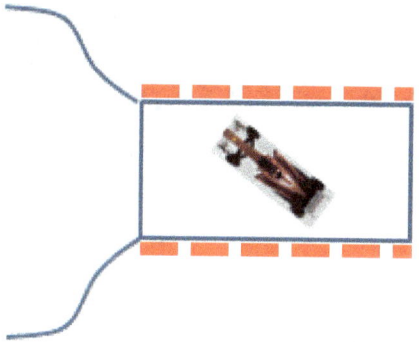

As an example, considering values referring to (old specs) F1 car we have:

Cz ≈ 2 with Cy ≈ 0.15 at 4° yaw.

Because tire friction coefficient is of the order of 1.8 and with a racing weight of 600Kg, the ratio of lateral tire force to aerodynamic lateral force at 180 km/h is approximately:

$$F_{z\,aero} \sim F_{z\,weight}$$
$$F_{y\,aero} \sim \frac{0.15}{2} F_{z\,aero} = 0.075 F_{z\,aero}$$

$$\frac{F_{y\ tire}}{F_{y\ aero}} > \frac{\left(1_{F_z\ aero} + 1_{F_z\ weight}\right) \cdot 1.8}{0.075_{F_z\ aero}} = 26$$

If not optimized, downforce in yaw (F1 car) may well reduce by more than 10%.
Very important note:

Today, races are won on corners.

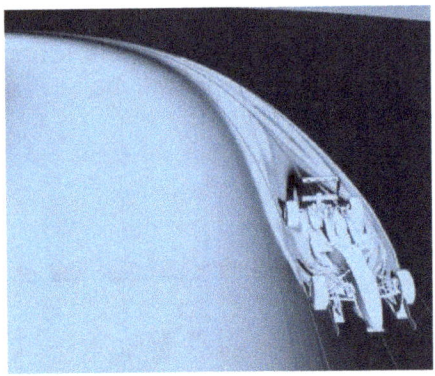

You need to do a lot of emphasis on the issue of knowing aerodynamic data of the car in yaw angles. Keep in mind the following thought:
In aerodynamics of competition, many efforts of all kinds to get several things are done, but the important ones are:

- Place the center of pressure where has to be ....
- Make the car has a lot of downforce and low drag.

Why try to have a lot of downforce?

In order for having a lot of grip or friction with asphalt in fast corners and to achieve faster cornering. This is the essence.

In curve it is then extremely "rare" that air comes parallel to the direction of the car; You will always have a yaw with respect to the path of the car; This yaw values can reach even 90 degrees.

Therefore, if you want the car to be high speed cornering, the car's aerodynamics must be studied when the air flow is directed not frontally; ie the car in yaw performance.

A few years ago this type of aerodynamic study was more than difficult; today, with existing wind tunnels and especially with CFD techniques it is difficult, but possible.

FLOW VISUALIZATION

Keep in mind that the air is not seen and in many cases, is necessary and appropriate to see what happens near a surface and on the same surface.

Sometimes is enough to know the values of downforce, the drag or pressure; but for overall design of a car or at least pre and manufacturing phases, it is convenient to visualize the flow to know "more" things that we can say from the measurable aerodynamic values.

3 activities that can be performed either in track or in wind tunnel are:

- Display flow "on" a surface.
- Display flow in an area.
- Measuring point velocities.

ON SURFACES

We can "paint" or "smear" the surface to study with some viscous substance; normally is used paraffin or oil derivatives; thus dust sticking gives references on how air flows "over" the surface to study.

Myself, to cheapen a test, I used "butter"
....

Using CFD is possible to visualize this flow that sticks to the surface; in most codes, there is a possibility of "oil flow"; if we try to display the speed, we see that gives us the value "0"; It is because the condition of wall is just that: zero speed:

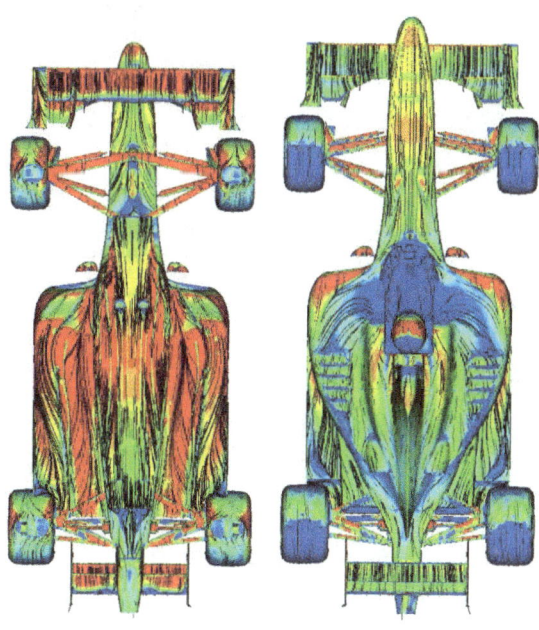

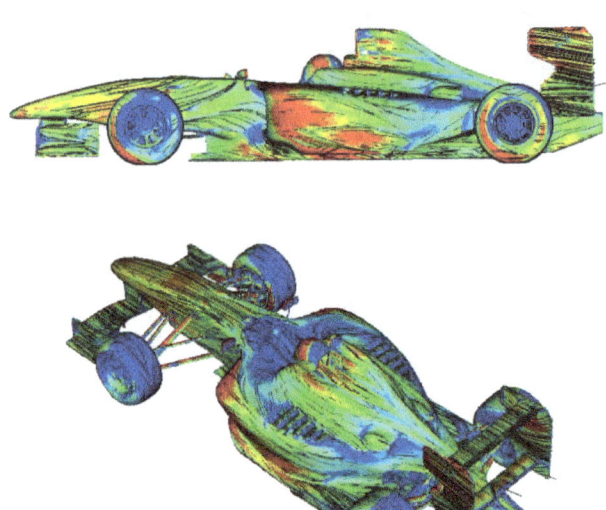

We've all seen medical x-rays, but surely few of us are able to "see something" in them; We say this, since flow view is one thing, but knowing what happens is another; from these views, we can know:

- The transition from laminar to turbulent flow,
- The separation of the boundary layer or the flow.
- Areas of low pressure (bulbs).
- Areas of high pressure.
- Etc ....

Let see some real examples:

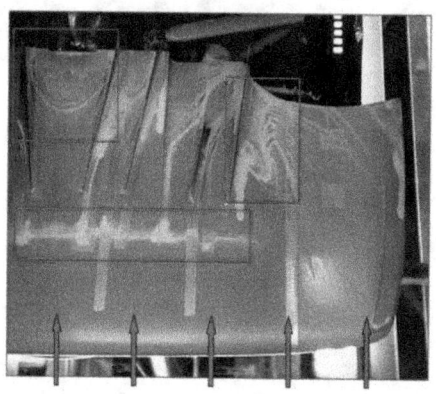

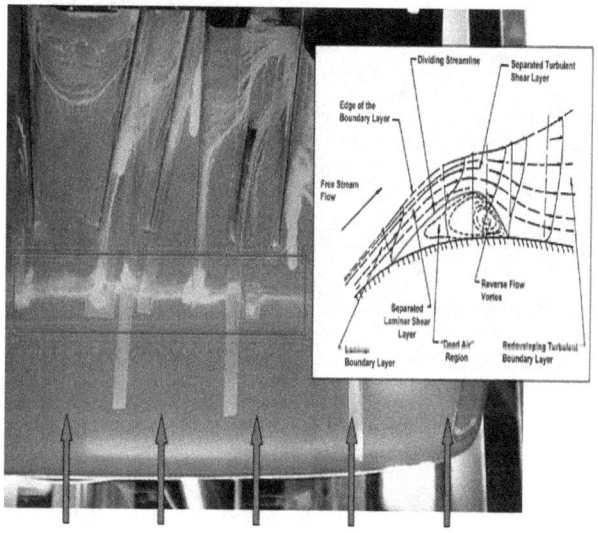

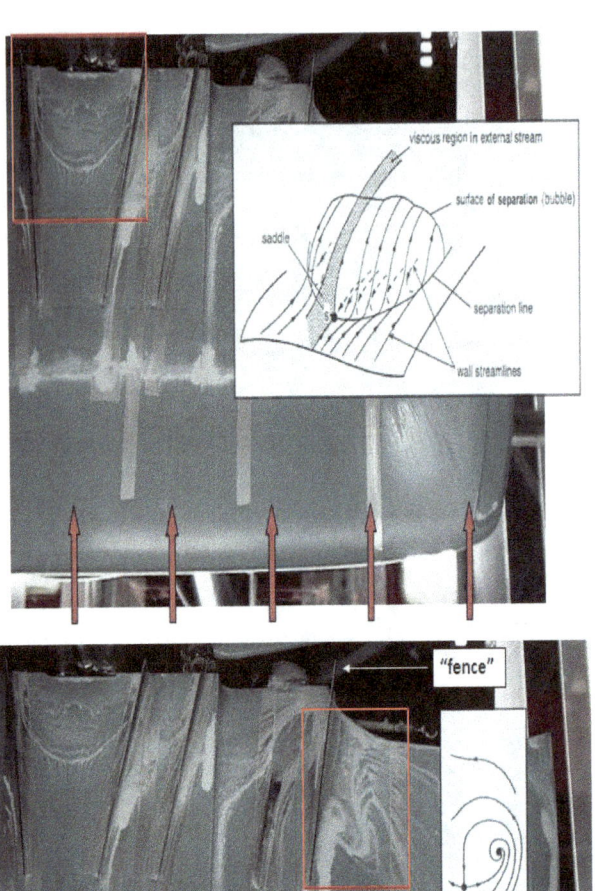

In the next picture we see that the flow below the rear wing will not peel; this means that the wing is operating correctly (is attatched):

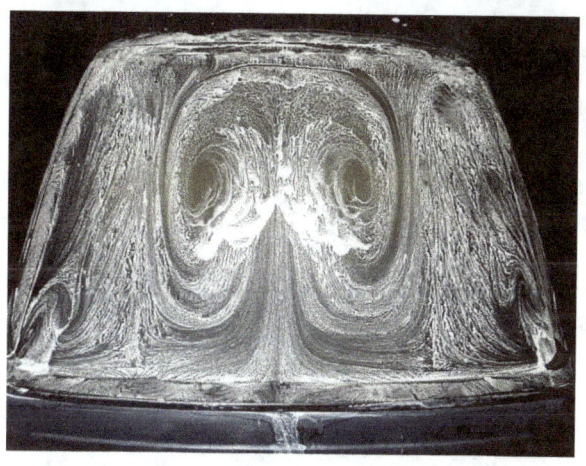

Example about ground and his operation:
   Bad: separation existing:

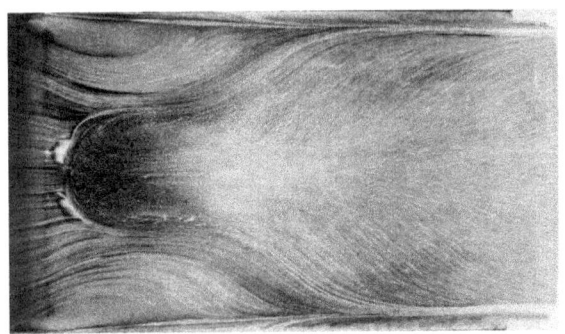

Good: simetrical:

Bad: separation and chaotic dynamic existing:

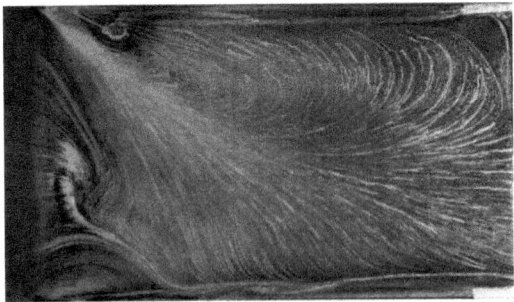

As a example for seeing the similitude between CFD simulation and reality, we can see the next image:

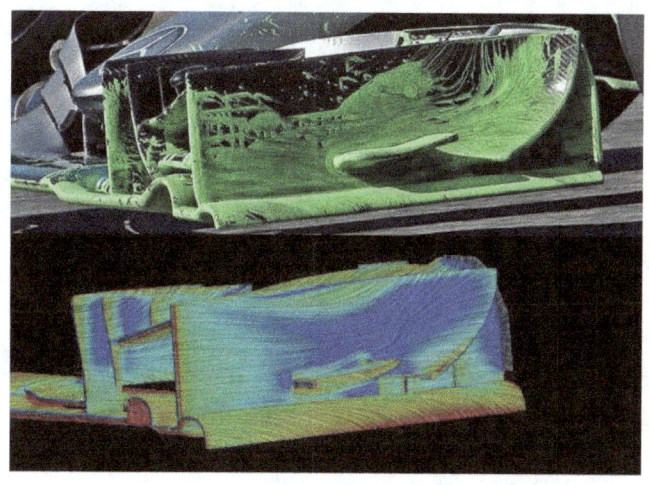

Another way of knowing flow on a surface is by placing sensors on said surface, measuring the pressure.

There is another way to visualize the flow on a surface; it is placing wool threads on the surface:

We see that the item of study is placed ahead of the car and above, so that in this way, the study is not affected by the air that the car would deflect.

Finally, as more cheaply and revealing one could say, we can replace the surface to study for "**fabric**"; observing said fabric, if bulges, low pressure exists, and if it sticks to the car say, there will be high pressure; that easy. Through this system we can practice a lot of aerodynamic driving on the road: we just have to put us next to a truck with canvas walls and watch; is incredible **the amount of** areas of low and high pressure we can observe.

FLOW IN AN AREA/VOLUME

This is visualizing the flow by throwing smoke; we can see the formation of turbulence, among other things:

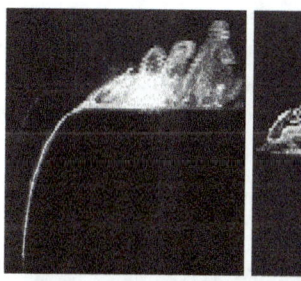
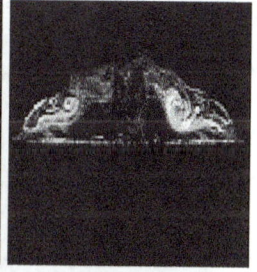

MEASUREMENT OF SPEED AND PRESSURE IN AN AREA/VOLUME

Pressure is the mother of all the properties of an airflow; from it you can get the rest (speed, etc ....); for this reason, when it comes to "know" how air flows in an area, it is necessary to measure the specific pressures in that area. For this, special systems that each team designs depending on what area you want to study are used; but in essence, they are punctual pressure sensors. Typically differential pressure sensors are used, because we are interested in knowing the pressure differences between pairs of points.

We work with the so called "pressure combs" that are nothing more than pipe racks, in which the point pressure is measured:

We may also need to measure pressures in other places, perhaps not as accessible, as the bottom of the car:

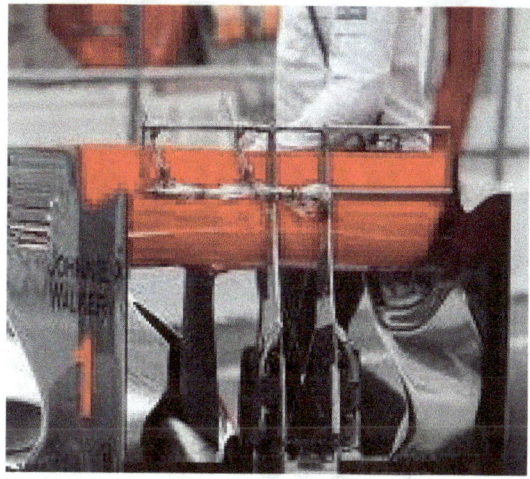

Adding these "combs", entails a loss of information rather large, since the comb itself interferes and greatly alters the airflow in that area; therefore it is necessary to place optimally these sensors, in order not to modify the flow excess.

It is much more useful to use CFD as it does not alter in any way the flow.

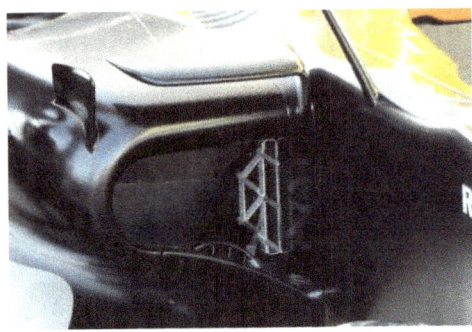

On a much larger scale, pressure combs are also used (this time with three-dimensional hot wire sensors) to analyze flows and turbulence that leaves a trail of plane:

Sometimes one needs to know the vibrations or deflections-forces that a certain element performes, as well as to know the downforce generated. In this case the device that they have invented to measure both is: if the sensor or sensors placed at the end of the bars are fixed to the front wing, inside the nose or inside of a support for this purpose that may contains the senso.

VISUALIZATION OF "HOT" AIR FLOW

We saw at the time, the extreme importance it has had in recent seasons "everything" related to exhausts. The stream of hot air coming out of the exhaust at high speed, is used from a streamlined view for many things, basically to issues directly related to the generation of downforce.

Furthermore, and not least, the air that cools the brakes which in turn are used for other aerodynamic effects are also exploited.

Knowing the dynamics of these gases, know where they flow, know where they end, and even know the evolution of the temperature, it has become today, essential. CFD simulations are capable of solving these problems.

If we have the car on the track and we want to know the evolution of hot air, we can use infrared camera to visualize and detect hot air.
We can look at certain areas:

- Difusser area.
- Sidepods' back area.
- Brakes and wheels.
- Exhausts.
- Etc.

Through colors, we can know the different temperatures, of both air and surfaces and parts:

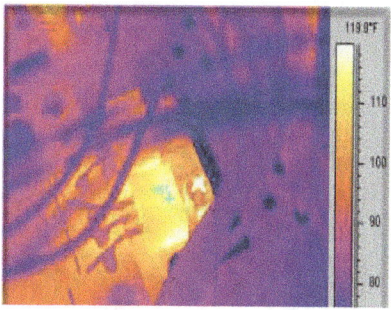

WIND TUNNEL SAMPLE

Sample Wind Tunnel (NASA courtesy):

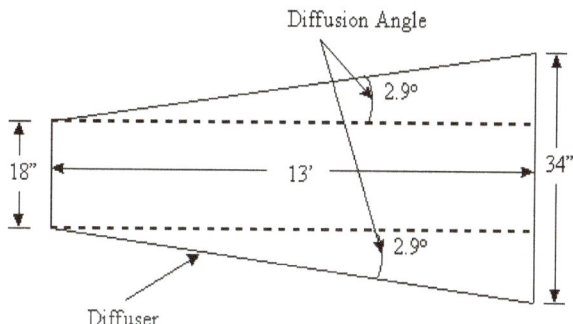

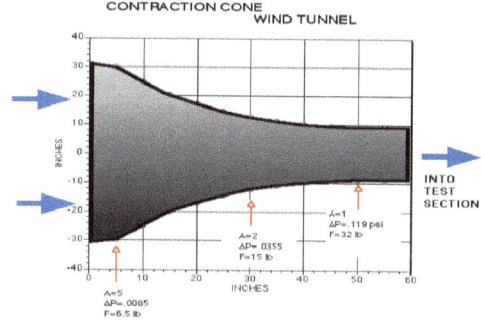

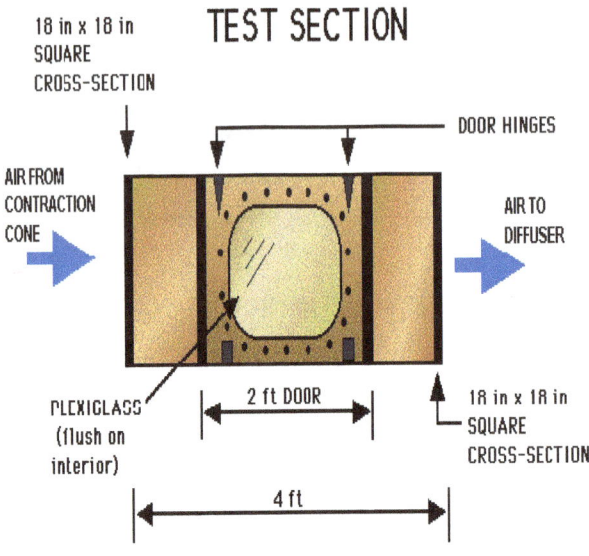

TEST SECTION

18 in x 18 in SQUARE CROSS-SECTION

DOOR HINGES

AIR FROM CONTRACTION CONE

AIR TO DIFFUSER

PLEXIGLASS (flush on interior)

2 ft DOOR

18 in x 18 in SQUARE CROSS-SECTION

4 ft

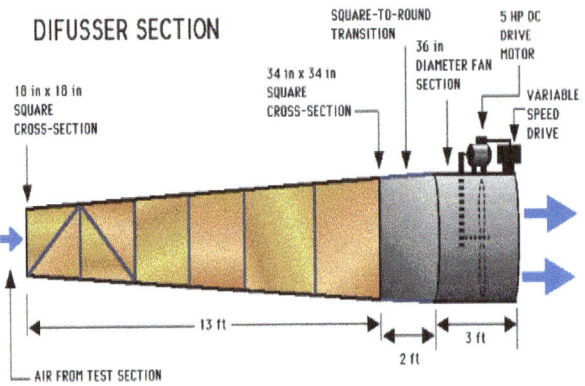

DIFUSSER SECTION

SQUARE-TO-ROUND TRANSITION

5 HP DC DRIVE MOTOR

36 in DIAMETER FAN SECTION

34 in x 34 in SQUARE CROSS-SECTION

18 in x 18 in SQUARE CROSS-SECTION

VARIABLE SPEED DRIVE

13 ft

2 ft

3 ft

AIR FROM TEST SECTION

90 mph speed maximum.

A design team of fifteen students was assembled at General B. O. Davis Jr. Aviation High School in May of 1995. These students went on a field trip to NASA Lewis on June 5, 1995, for a wind tunnel project kickoff meeting. The students received a tour of the various wind tunnels at NASA Lewis in the morning. In the afternoon, the students were briefed by a NASA engineer, about basic aeronautical theory and about the parts of a wind tunnel.

The students began the design of their tunnel once back at school. Two NASA Engineers participated in the project by going to the school to review the students' design and to give suggestions. Once the design was completed, the students began building the various parts of the tunnel. The construction of the tunnel was completed during the summer of 1995. The students volunteered to work on the construction of the tunnel during the summer.

The wind tunnel contains three main parts, the contraction cone at the front of the tunnel, the test section in the middle of the tunnel, and the diffuser at the back of the tunnel. The project team wanted to create the largest tunnel that their budget would allow. They, therefore, iterated between the size of the tunnel and the cost associated with it, to determine how big they could make the tunnel.

The size of test section and the size of the contraction cone are related through a contraction ratio. The contraction ratio used, was 12 to 1. This is the ratio between the area of the contraction cone and the area of the test section. It was determined that a 5' x 5' contraction cone was the maximum size that could be built. This then dictated that the test section would be 1.5' x 1.5'. The shape of the contraction cone was designed by the NASA Lewis aeronautical engineers working on the project.

They developed a bell curve shape that would cause laminar flow and eliminate flow separation. It was determined that an optimal maximum speed for the tunnel would be around 90 mph.

The next step was to find the volumetric flow rate of a tunnel with a 1.5' x 1.5' test section and a velocity of 90 mph. Ninety miles an hour was converted to 132 ft/sec. Next, the area of the test section was found to be 2.25 square feet. The volumetric flow rate equation is Q = v x A. So, Q = 132 ft/sec x 2.25 ft squared = 297 cubic ft/sec. This converted into cubic ft/min, is 17,820 CFM. The size of the motor was based on the desired CFM output of the fan. With a 5 hp motor, a 36" diameter fan can move 20,650 cubic feet of air per minute, this was greater than the desired CFM output, so these two pieces of equipment were selected.

The length of the diffuser section was dictated by the size of the fan, which was selected to be 36" in diameter. A square to round conversion piece was designed and the square portion turned out to be 34" x 34". This became the dimensions of the end of the diffuser section. The diffusion angle was kept below 6 degrees. The diffusion angle ended up being 2.9 degrees on each side for a total of a 5.8 degree diffusion angle. This made the diffuser section 13' length provides further explanation about the diffusion angle and the diffuser section.

WIND TUNNEL TEST PLUNGING (HEAVE) AND PITCH VIBRATION

With this tunnel, is possible to vibrate the wing or any object, in order to know the variation form example, of downforce. This system is ideal to applicate to chapter number 17 (aero post rig analisys).

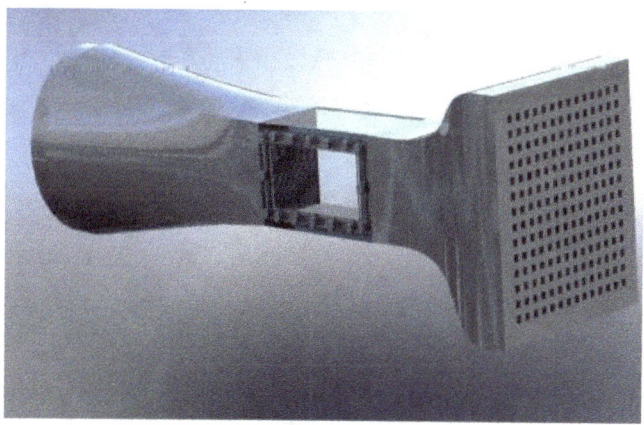

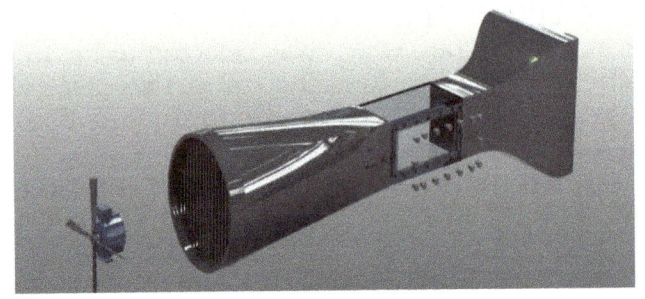

Pitch Vibration:

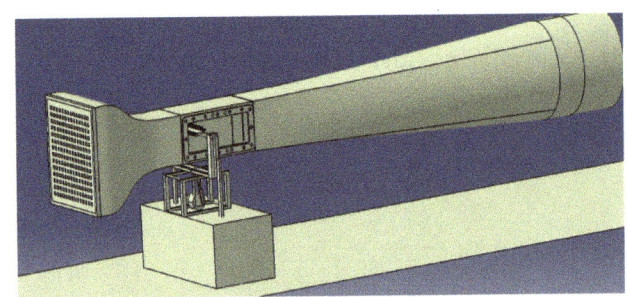

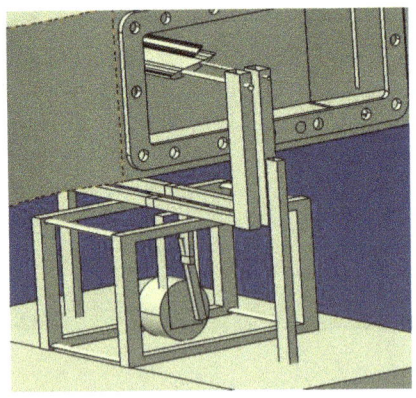

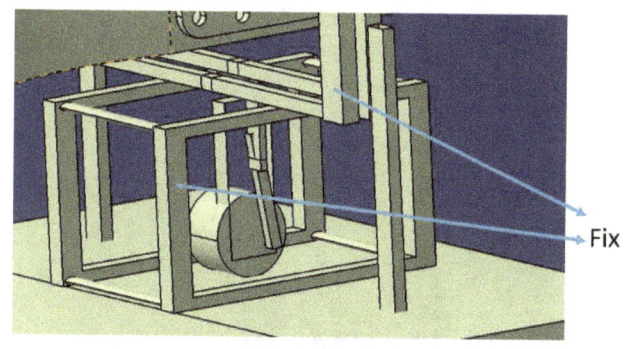

Fix

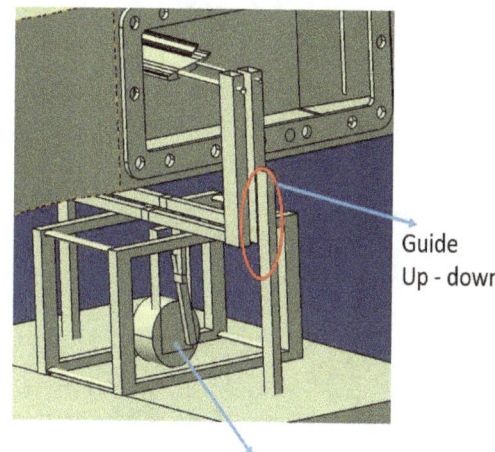

Guide
Up - down

Engine + biela

Heave Vibration:

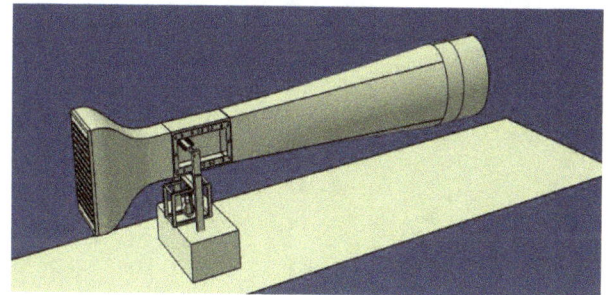

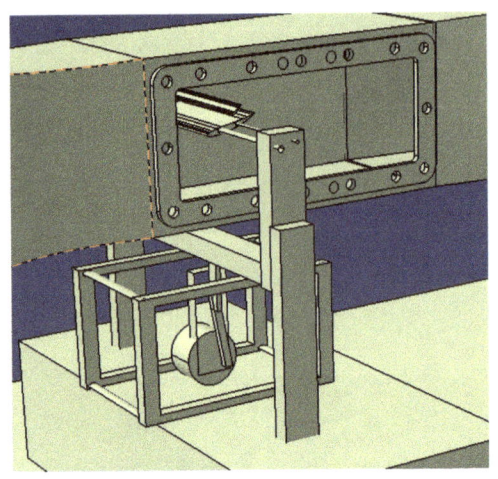

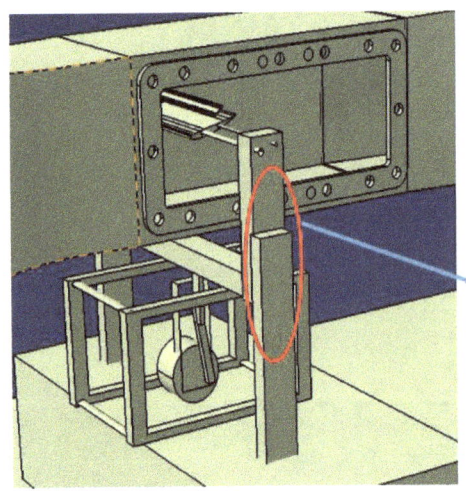

Guide
Up - down

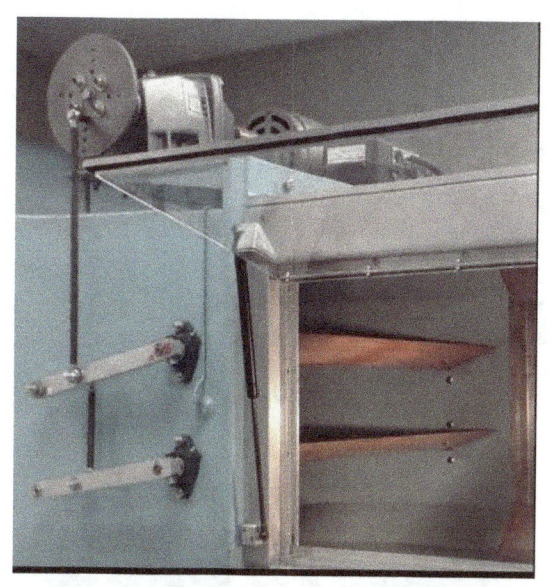

Quarter Car model:

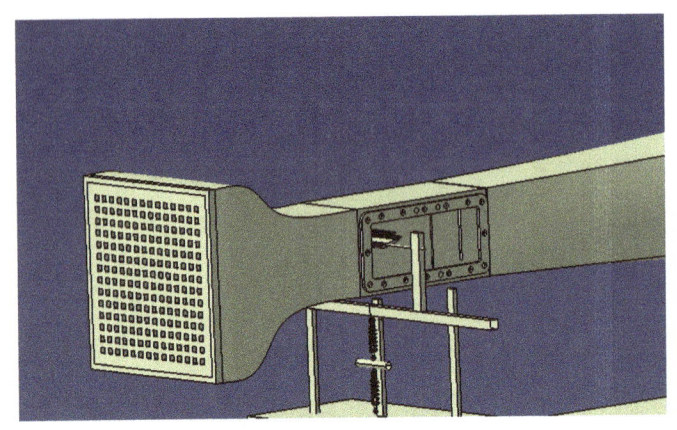

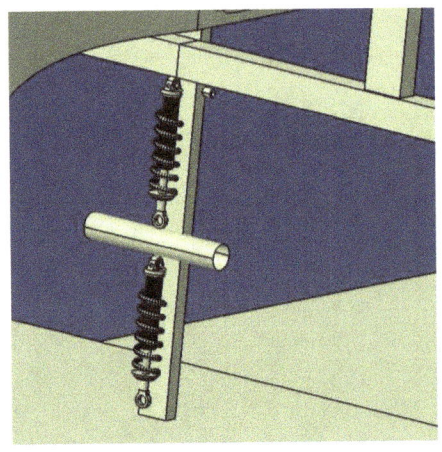

WIND TUNNEL SAMPLE – ASPIRATION

The contraction duct:

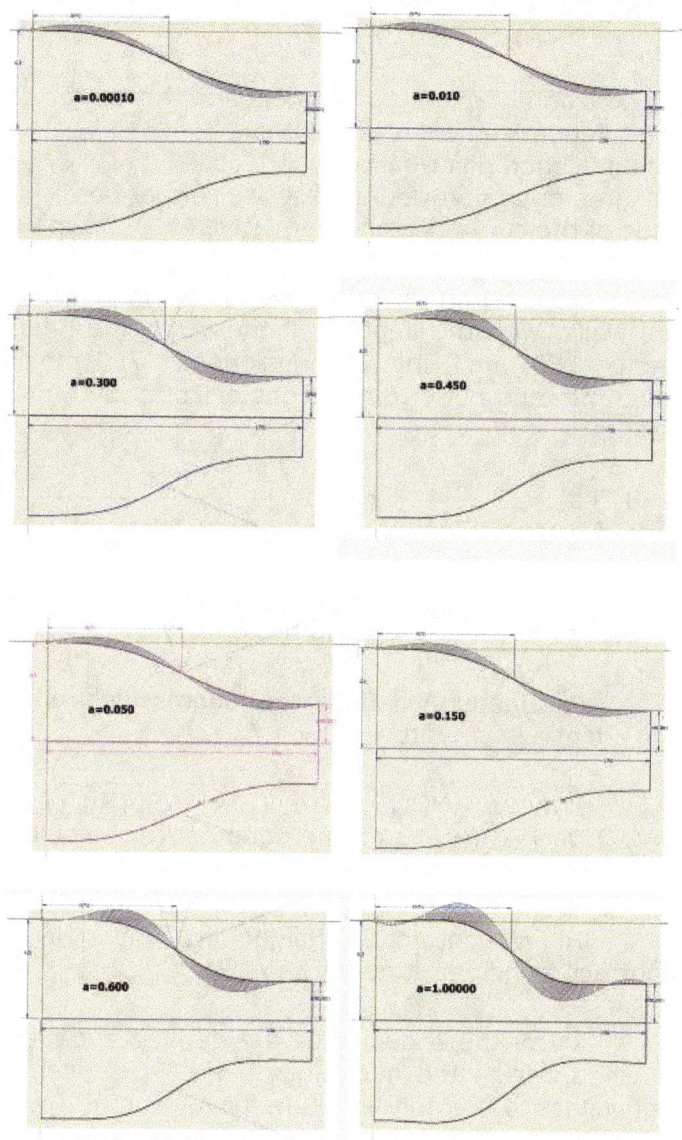

The geometry of the contraction duct is to be defined by the following constraints:

- a horizontal tangent at both ends of the curve (first derivative equals zero here).
- an inflection point half way.
- degree of curvature decreases to zero at both ends of the curve (second derivative equals zero).

Deriving an equation for half of the curve of the walls, with the depth L of the contraction duct being 1750 mm , Inlet dimensions of 1750 mm x 1250 mm and outlet dimensions of 700 x 500 mm, results in:

$X=875*t$
$Y=875*(A*t^2-7*A*t/3 +B*t +4*A/3 -2*B)*(t^3)$

With B= 0.3 for the side walls and B=0.214286 for the top and bottom walls.

Parameter A defines the 'aggressiveness' of the change in curvature.

Provides $Y(X)=$ Hi- (Hi- He)$6(X')^5$ – $15(X')^4 + 10(X')^3$ as the equation that resulted in the best performing contraction duct,with:

Hi = contraction height at inlet   He = contraction height at outlet and L/Hi > 0.89

It can be shown that the derived equation, used to design the geometry in the 3D-model, is equivalent to the Bell-equation for a value of A = ~ 0.000001

The contraction duct is the largest single part of the wind tunnel, weighing in at 320kg. The part is also designed to be 10mm smaller and 100mm lower than the door of the wind tunnel room.

The outer frame, composed from 6mm lasercut steel, was assembled first and tack welded. This structure defined the geometry of the contraction duct.

The 2mm wall panels were then positioned inside the structure, bent outward against the frame and tack welded. Looking down into the upright contraction duct. The plywood slats are being used to push the wall sheets outwards against the frame.

MESURES

Basically in a wind tunnel, is necessary to calculate the:
- Downforce rear and front wheels axes.
- Drag Total.
- Moments.
- Etc...

This etc, that is the vibrations, etc...
In order to calculate the forces, it used the load cells:

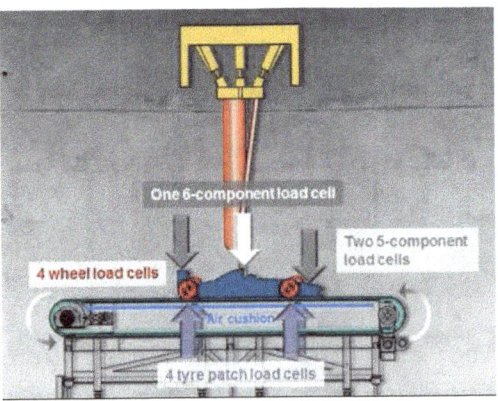

The best method (less drag induced in car) for install and measure the forces and moments, is from up (vertical bar):

Also is possible to install car and wheels "together or in contact" or not.... That, is complicate to resolve.

WIND TUNNEL TEST – CORRELATION WITH REALITY

For an F1 racing car:

- 3 different kinds of modifications to the "baseline" geometry:

    - *Top rear wing*

    - *"Barge Boards"*

    - *Front wing endplates*

- Tested in 2 different WT using similar 50% scale model.

- Computed in CFD to <u>simulate WT</u> results (i.e. <u>model scale</u>, and different WT geometry included in the simulation).

For an LMP racing car:

- 1 alternative configuration of the whole front bodywork.

- Tested in 1 WT using 40% scale model.

- Computed in CFD to <u>simulate free-air</u>, <u>full-</u>

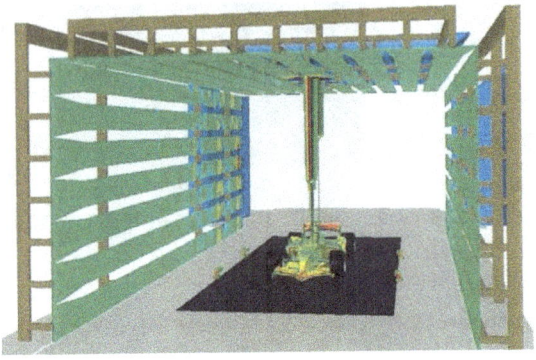

F1 racing car:

- TRW, WT1
- TRW, WT2
- BB, WT1
- FW, WT1
- FW, WT2

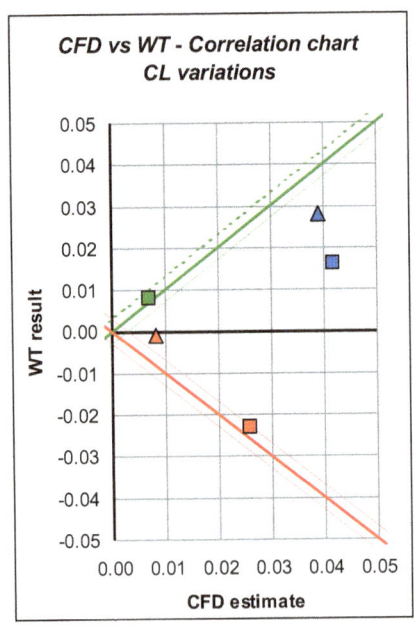

CFD vs WT - Correlation chart
CL variations

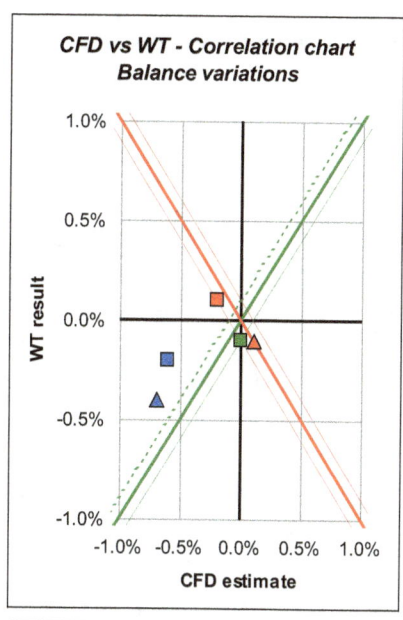

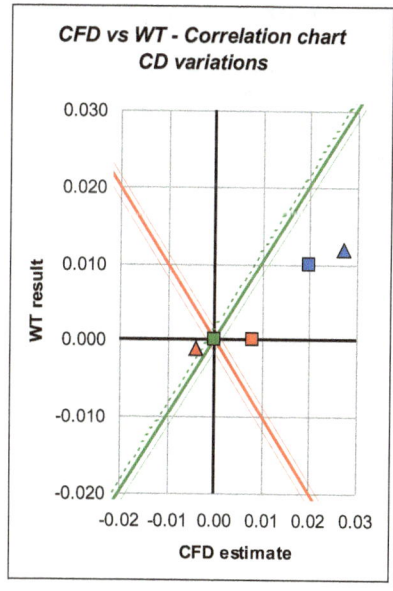

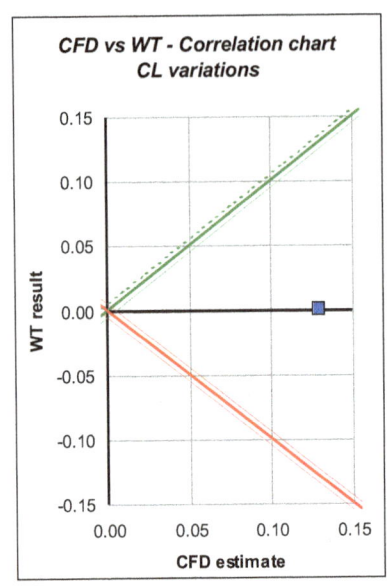

LMP racing car:

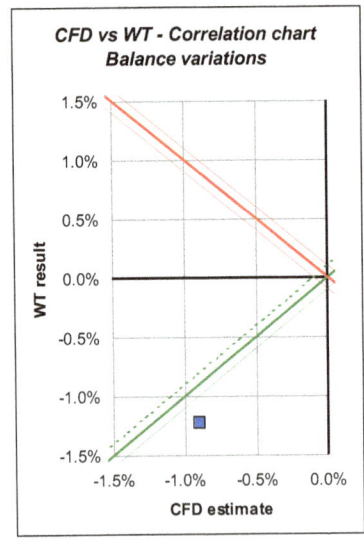

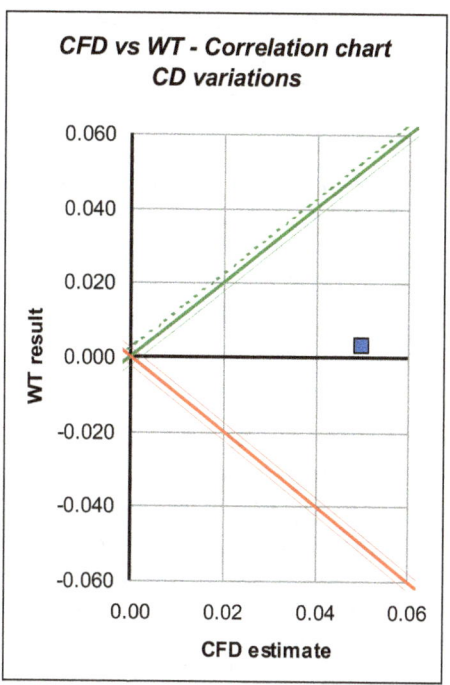

CFD – THEORY

Aim of the chapter

The aim of the present chapter is to provide a solid background in the theory behind the computational fluid mechanics methods (CFD) as long as a set of best practice guidelines in order to approach some of the typical simulations carried in the motorsport field.

The chapter is directed to a novice in the field of CFD. There are great resources like [1], [2] and [3], just to mention a few, that provide a deeper knowledge of the particularities of a field as extensive as CFD. However, the author's intention is providing a painless start point, keeping a motorsport focus. If the reader would need a deeper knowledge of the points treated in this chapter, he should not worry as suitable bibliography is being provided throughout the chapter.

From the governing equations to final results, a theory background

In order to approach the CFD world, there must be no short cuts and the proper first step should be having a good understanding of the equations beneath the physical phenomena. In section 0, the governing equations will be introduced and the physical meaning of the different terms in them will be discussed. Once a proper background of the mathematical modelling had been given, the methods to convert this mathematical model into a simpler, algebraic model will be approached in section 0. The way non-linearities that may arise are treated will be discussed in section 0. Finally, the most typical algorithms for solving the fluid flows will be explained in section 0.

The concepts provided in this chapter will be supported by several easy-to-follow Matlab examples that the author believes will help to settle the previous introduced concepts. These examples are not intended to be efficiently programmed codes but to provide even the unexperienced programmer a structure of how the different techniques are implemented.

The governing equations

In order to derive the governing equations, that as we mentioned will be the pillars of the ongoing discussion, it is necessary to adopt a mathematical model to describe the fluid. In this sense, there are several approaches, the most typical are cited bellow:

1. Considering a Fluid Volume fixed in space.
2. Considering a Fluid Volume moving with the fluid.
3. Dividing the domain in infinitesimal parts fixed in space.
4. Consider the infinitesimal parts moving with the fluid.

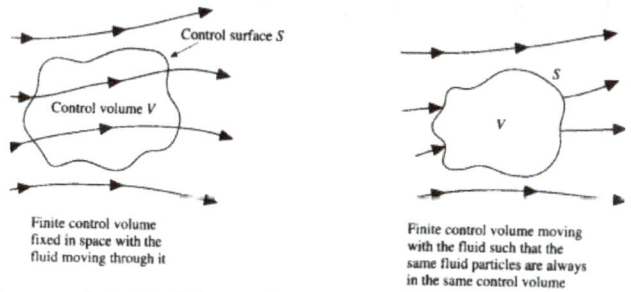

Figure 0.1 Fluid models describing a (left) fixed control volume and (right) a moving control volume

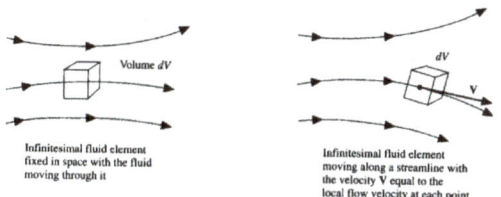

Figure 0.2 Fluid models describing a (left) fixed infinitesimal element and (right) a moving infinitesimal element

We will focus on the approach designated as 1, a fixed control volume, in that way we will retrieve the governing equations in their integral conservative form. This form is the one used by the vast majority of CFD software which are mainly based on finite volume discretization techniques. For a complete derivation of the governing equations using any of the fluid models mentioned, refer to [1].

Before attempting to derive the governing equations, an important concept must be known by the reader: the substantial derivative.

The substantial derivative

In fluid mechanics there are mainly two approaches to describe the fluid flow: the lagrangian approach and the eulerian approach.

The lagrangian approach put its focus on an infinitesimal particle of fluid moving within the flow. In this approach we describe the position (x_P), velocity (u_P), density (ρ_P) and any other macroscopic property (ϕ_P) as:

Equation 0-1

$$x_P = x_p(t)$$

Equation 0-2

$$u_P = u_p(t)$$

Equation 0-3

$$\rho_P = \rho_P(t)$$

Equation 0-4

$$\phi_P = \phi_P(t)$$

Where x_i represents the position of the infinitesimal fluid element P.

If we would like to calculate the rate of change of the property ϕ from two time instants, t and $t + \Delta t$, we could approximate it as:

$$\frac{\phi_P(t + \Delta t) - \phi_P(t)}{\Delta t}$$

In the limit:

$$\lim_{\Delta t \to 0} \frac{\phi_P(x_i(t + \Delta t), t + \Delta t) - \phi_P(x_i(t), t)}{\Delta t} = \frac{D\phi}{dt}$$

That is the so-called substantial derivative.

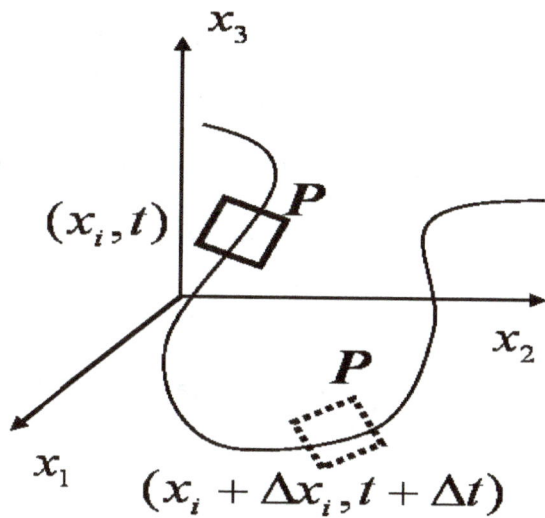

Figure 0.3 Representation of an infinitesimal fluid element in two different time instants

     The eulerian approach is completely different. In order to understand it, imagine yourself in a wind tunnel with a pitot tube. If you wanted to measure fluid velocity, you would place the pitot tube in the desired location and register the velocity. If you may want to describe the motion from a lagrangian point of view, you should move with the same velocity of an infinitesimal element during your record of velocity. It is clear at this point the importance of the eulerian approach as it is the one employed while performing fluid flow experiments. From an eulerian point of view, the velocity (u_P), density (ρ_P) and any other macroscopic property (ϕ_P) are described as functions of both time and space:

Equation 0-5

$$u = u(x,t)$$

Equation 0-6

$$\rho = \rho(x,t)$$

Equation 0-7

$$\phi = \phi(x,t)$$

This approach is the one is going to be used from now on. However, the fundamental principles from which the governing equations will be derived refer to substantial derivatives that are immediate with the lagrangian approach but not so with the eulerian. This issue is solved applying the Reynolds transport theorem that states the following:

Equation 0-8

$$\frac{D}{dt} = \frac{\partial}{\partial t} + \boldsymbol{u} \cdot \nabla$$

That can also be referred in integral form with the help of Gauss's theorem as:

Equation 0-9

$$\frac{D}{dt} \int\int\int_V \boldsymbol{f} \, dV = \frac{\partial}{\partial t} \int\int\int_V \boldsymbol{f} \, dV + \int\int_S \boldsymbol{f}(\boldsymbol{u} \cdot \boldsymbol{n}) dS$$

Being f scalar, vector or tensor valued.

If any doubt still exists in the reader, it will be clear out after examining the following simple example. Imagine, for the sake of simplicity, we have a fluid moving uniformly at a velocity $u_0 = 5\,m/s$ from a zone A at 30°C to a zone B at 20°C, as represented in **Error! Reference source not found.**. This is a one dimensional steady example and has no physical meaning, its only purpose is helping the reader to fully understand the concepts presented.

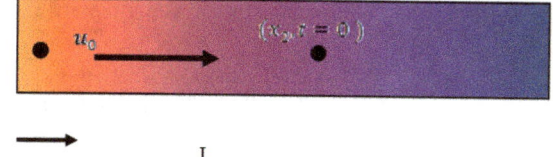

x

L

Figure 12.4 Easy example to illustrate the differences between eulerian and lagrangian approach.

It is obvious that any particle cools down during its movement. From a lagrangian point of view, we describe the particle noted as 1 as:

Equation 0-10
$$x_1(t) = x_1(0) + u_0 t$$

Equation 0-11
$$u_1(t) = u_0$$

Equation 0-12
$$T_1(t) = T_A + \frac{x_1(t) - x_1(0)}{L}(T_B - T_A) = T_A + \frac{u_0 t}{L}(T_B - T_A)$$

In the same way, we would have a different set of expressions for an element starting in 2 and so on for any other element in the fluid. The calculation of the rate of change of the temperature is straightforward:

$$\frac{DT_1}{dt} = \frac{u_0}{L}(T_B - T_A)$$

Note that in this easy example, this rate will be equal for any fluid element.

Using a eulerian approach, we could simply describe the fluid field as:

Equation 0-13
$$u(x, t) = u_0$$

Equation 0-14
$$T(x, t) = T_A + \frac{x}{L}(T_B - T_A)$$

In order to retrieve the substantial derivative, we use the expression of Equation 0-8:

Equation 0-15

$$\frac{D}{dt}T = \frac{\partial}{\partial t}T + \boldsymbol{u} \cdot \nabla T = \underbrace{0(\text{steady})}_{local\ derivative\ term} + \underbrace{u_0 \frac{T_B - T_A}{L}}_{convective\ term}$$

$$= \frac{u_0}{L}(T_B - T_A)$$

At this point we can observe several important points:

- The fluid flow described is steady; it does not vary with time. With the eulerian approach with can describe it naturally without including time in our expressions; however, as the lagrangian approach follows every element and they are moving, time must be incorporated to the lagrangian expressions.
- In the eulerian approach we can distinguish two terms in the substantial derivative: the local derivate and the convective term. The local derivative accounts for the effects of the time local variations of the property while the convective term accounts for the variations due to the movement of the flow to zones with different values of the property (here is where the gradient plays its role).

With these concepts clear in mind, we can move forward and face the derivation of the governing equations.

The continuity equation

All the governing equations are derived from conservation principles and the continuity equation is not an exception. The principle that sustained the continuity equation is widely known and states that *mass is conserved*. If we now apply it to our preferred fluid model, a fixed in space volume control, we can express the total mass of the control volume as:

Equation 0-16

$$M_{control\ volume} = \int \int \int_V \rho\, dV$$

Where the control volume mass has been expressed in terms of a volume integral of the density throughout V, our fixed control volume

Then it is straightforward to deduce the time rate of change of this quantity:

Equation 0-17

$$\frac{\partial}{\partial t} \int \int \int_V \rho\, dV$$

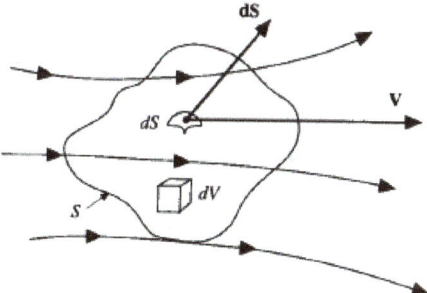

Figure 0.4 Explanation of the mass flux over a control volume

Now the only thing remaining is expressing the net mass flux over the surface of our control volume. This can be expressed as a surface integral of the density multiplied by the velocity components normal to the boundary of our control volume, in other words the scalar product $\boldsymbol{u} \cdot \boldsymbol{n}$ being $\boldsymbol{n}$ the nomal versor (unitary vector) of the boundary in each point:

Equation 0-18

$$Net\ flux = \int \int_S \rho \boldsymbol{u} \cdot \boldsymbol{n}\, dS$$

If we take $\boldsymbol{n}$ pointing to the outside of our control volume, this flux will be positive when mass is abandoning the control volume. So having this in mind, we can put Equation 0-16 and Equation 0-18 together:

$$Rate\ of\ mass\ change\ in\ the\ control\ volume$$
$$= -Net\ Flux\ abandoning\ the\ control\ volume$$

$$\frac{\partial}{\partial t} \int \int \int_V \rho\, dV = - \int \int_S \rho \boldsymbol{u} \cdot \boldsymbol{n}\, dS$$

$$\Rightarrow \frac{\partial}{\partial t} \int \int \int_V \rho\, dV + \int \int_S \rho \boldsymbol{u} \cdot \boldsymbol{n}\, dS = 0$$

Equation 0-19

$$\frac{\partial}{\partial t} \int \int \int_V \rho\, dV + \int \int_S \rho \boldsymbol{u} \cdot \boldsymbol{n}\, dS = 0$$

The momentum equation

The principle from which we are going to derive the momentum equation is the well-known Newton's second law that states that the rate of change of the linear momentum ($\boldsymbol{p} = m\boldsymbol{u}$) is equal to forces applied. The linear momentum of our control volume can be expressed as:

Equation 0-20

$$p_{control\ volume} = \int \int \int_V \rho u\, dV$$

Newton's second law can be expressed in terms of an equation as:

Equation 0-21

$$\frac{D}{dt}p = F$$

Now we can apply the Reynolds's transport theorem to the left side term obtaining the following:

Equation 0-22

$$\frac{D}{dt}p = \frac{\partial}{\partial t}\int \int \int_V \rho u\, dV + \int \int_S \rho u(u \cdot n)\, dS$$

Now it is time to put our attention on the different forces applied over our control volume. We can decompose them in two big groups: volumetric forces and superficial forces. Volumetric forces are forces that act all over the domain, a clear example is gravity. We can express them as:

$$F_v = \int \int \int_V f_v dV$$

In the case of forces due to gravity, we can express them with $f_v = \rho g$, being g the gravitational acceleration.

The other important forces that are being considered, the superficial forces, can be described employing the mathematical description of the strain rate tensor. We can note the strain rate tensor as:

$$\bar{\bar{\tau}}(x,t) = \begin{pmatrix} \tau_{xx}(x,t) & \tau_{xy}(x,t) & \tau_{xz}(x,t) \\ \tau_{yx}(x,t) & \tau_{yy}(x,t) & \tau_{yz}(x,t) \\ \tau_{zx}(x,t) & \tau_{zy}(x,t) & \tau_{zz}(x,t) \end{pmatrix}$$

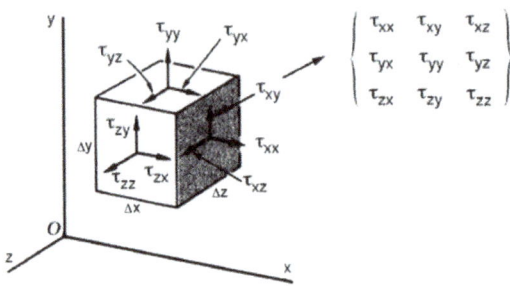

Figure 0.5 Graphical representation of the strain rate tensor

In this way, we can obtain the force applied to a surface as:

$$F_S = \int\int_S \bar{\bar{\tau}} \, n \, dS$$

If we know shift out focus to the physics beneath this strain rate tensor, we can distinguish two fundamental types of superficial forces. One of them are forces exerted by the pressure surrounding our control volume, the other type of superficial forces are caused by the viscous friction between fluid particles when in motion. As the pressure forces are always normal to the surface where they are applied, we can decompose our strain rate tensor as follows:

$$\bar{\bar{\tau}} = PI + \bar{\bar{\tau}}^* = \begin{pmatrix} P(x,t) & 0 & 0 \\ 0 & P(x,t) & 0 \\ 0 & 0 & P(x,t) \end{pmatrix}$$

$$+ \begin{pmatrix} \tau_{xx}^*(x,t) & \tau_{xy}^*(x,t) & \tau_{xz}^*(x,t) \\ \tau_{yx}^*(x,t) & \tau_{yy}^*(x,t) & \tau_{yz}^*(x,t) \\ \tau_{zx}^*(x,t) & \tau_{zy}^*(x,t) & \tau_{zz}^*(x,t) \end{pmatrix}$$

Being I, the Identity matrix.

Now, it is time to put the part of the strain rate tensor due to fluid's viscosity in terms of the flow unknowns $(\rho, u, P, T \dots)$, a constitutive law is needed that relates this term with the physical local characteristics of the flow. From now on we will consider the fluid as Newtonian and consequently, we will employ the constitutive law for Newtonian Flow, stated by Stokes in 1845:

$$\bar{\bar{\tau}}^*$$

$$= \begin{pmatrix} \lambda(\nabla \cdot \boldsymbol{u}) + 2\mu \frac{\partial u_x}{\partial x} & \mu\left[\frac{\partial u_y}{\partial x} + \frac{\partial u_x}{\partial y}\right] & \mu\left[\frac{\partial u_z}{\partial y} + \frac{\partial u_y}{\partial z}\right] \\ \mu\left[\frac{\partial u_x}{\partial y} + \frac{\partial u_y}{\partial x}\right] & \lambda(\nabla \cdot \boldsymbol{u}) + 2\mu \frac{\partial u_x}{\partial x} & \mu\left[\frac{\partial u_z}{\partial y} + \frac{\partial u_y}{\partial z}\right] \\ \mu\left[\frac{\partial u_x}{\partial z} + \frac{\partial u_z}{\partial x}\right] & \mu\left[\frac{\partial u_y}{\partial z} + \frac{\partial u_z}{\partial y}\right] & \lambda(\nabla \cdot \boldsymbol{u}) + 2\mu \frac{\partial u_x}{\partial x} \end{pmatrix}$$

Where μ, represents the molecular viscosity and λ, the volume viscosity (also known as bulk viscosity or second viscosity).

If we now put together all the terms of Equation 0-17 expanded, we obtain:

$$\frac{\partial}{\partial t} \int \int \int_V \rho u \, dV + \int \int_S \rho u(u \cdot n) \, dS$$

$$= \int \int_S \bar{\bar{\tau}} n \, dS + \int \int \int_V f_v dV$$

$$= \int \int_S PI \, n \, dS + \int \int_S \bar{\bar{\tau}}^* n \, dS + \int \int \int_V f_v dV$$

Rate of linear momentum change in the control volume
= Forces applied over the control volume

Equation 0-23

$$\frac{\partial}{\partial t} \int \int \int_V \rho u \, dV + \int \int_S \rho u (u \cdot n) \, dS$$

$$= \int \int_S PI n \, dS + \int \int_S \overline{\overline{\tau}} n \, dS + \int \int \int_V f_v dV$$

The energy equation

In order to derive the energy equation we are going to use exactly the same procedure as in the previous ones. We will formulate a physical conservation principle and explore how to express it in terms of mathematical equations applied to a fluid model of a fixed control volume. The physical principle is one that again is well known by the reader: *energy is conserved*. We express the energy contained in a control volume as:

$$Energy_{control \, volume} = \int \int \int_V \rho \left(e + \frac{1}{2} |u|^2 \right) dV$$

Where e, represents the specific internal energy of the fluid. As energy is conserved in a closed, the time variations of the energy inside our control volume can only be caused by energy exchanges with its surroundings and there are only two ways of doing so, exchanging heat and work:

$$\frac{D}{dt} \int \int \int_V \rho \left(e + \frac{1}{2} |u|^2 \right) dV$$
$$= Net \, heat \, flux$$
$$+ \, Rate \, of \, work \, done \, on \, the \, control \, volume$$

Now let's focus on the heat transfer between our control volume and the environment. This heat flux can be very varied and we can divide it in: volumetric heating (due to chemical reactions, radiation emission or other phenomena like the Joule effect) and heat transfer across the surfaces of the control volume (due to conduction).

We can express the volumetric heat in a similar fashion as we did with the volumetric forces:

$$Volumetric\ heat\ flux = \int\int\int_V \rho\dot{q}_v\,dV$$

Where $\dot{q}_v$ has IS units of W/kg. The concrete description depends of subjacent physical phenomena and it is out of focus at this point.

For describing the heat fluxes due to conduction we employ another constitutive law that will probably be known by the reader, the Fourier's law that states the following:

$$\dot{q} = -k\nabla T$$

With $\dot{q}$ in W/m^2 in IS units. The net flux over the control volume is expressed as:

$$Conduction\ heat\ flux = -\int\int_S k\nabla T \cdot \boldsymbol{n}\,dS$$

Summing all the heat fluxes:

$$Net\ heat\ flux = \int\int\int_V \rho\dot{q}_v\,dV - \int\int_S k\nabla T \cdot \boldsymbol{n}\,dS$$

Let's move on the rate of work done over the control volume. The main forces that can exert work over the control volume are, not surprisingly, the same reviewed during the derivation of the momentum equation: volumetric and superficial forces. The rate of work of a force acting over a moving element can be obtained as the scalar product of the force with the element's velocity, the same principle applies in fluids so we can express firstly the rate of work of the volumetric forces as:

$$\text{Rate of work of the volumetric forces } = \iiint_V f_v \cdot u \, dV$$

We can do exactly the same with superficial forces:

$$\text{Rate of work of the superficial forces}$$
$$= \iint_S (\bar{\bar{\tau}} n) \cdot u \, dS =$$
$$= \underbrace{\iint_S P(u \cdot n) \, dS}_{rate\ of\ work\ due\ to\ prssure\ forces}$$
$$+ \underbrace{\iint_S (\bar{\bar{\tau}}^* n) \cdot u \, dS}_{rate\ of\ work\ due\ to\ viscous\ forces}$$

If we now put all the pieces together:

$$\text{Rate of change of energy inside the CV}$$
$$= Net\ heat\ flux$$
$$+ Rate\ of\ work\ done\ on\ the\ CV$$

$$\frac{D}{dt} \iiint_V \rho \left(e + \frac{1}{2} |u|^2 \right) dV =$$
$$= \iiint_V \rho \dot{q}_v \, dV - \iint_S k \nabla T \cdot n \, dS$$
$$+ \iiint_V f_v \cdot u \, dV +$$
$$+ \underbrace{\iint_S P(u \cdot n) \, dS}_{rate\ of\ work\ due\ to\ prssure\ forces}$$
$$+ \underbrace{\iint_S (\bar{\bar{\tau}}^* n) \cdot u \, dS}_{rate\ of\ work\ due\ to\ viscous\ forces}$$

$$\frac{\partial}{\partial t}\iiint_V \rho\left(e+\frac{1}{2}|u|^2\right)dV + \iiint_V \rho u \cdot \nabla\left(e+\frac{1}{2}|u|^2\right)dV$$

$$=$$

$$= \iiint_V \rho \dot{q}_v\, dV - \iint_S k\nabla T \cdot n\, dS$$

$$+ \iiint_V f_v \cdot u\, dV + \iint_S P(u\cdot n)\, dS +$$

$$+ \iint_S (\overline{\overline{\tau}}^*\, n)\cdot u\, dS$$

This last expression is commonly used in liquids, in gases is more often to express it in terms of enthalpy rather than internal energy. Being $h = e + P/\rho$, we can express the energy equation as:

$$\frac{\partial}{\partial t}\iiint_V \rho\left(h+\frac{1}{2}|u|^2\right)dV + \iiint_V \rho u \cdot \nabla\left(h+\frac{1}{2}|u|^2\right)dV$$

$$=$$

$$= \frac{\partial}{\partial t}\iiint_V P\, dV$$

$$+ \iiint_V \rho \dot{q}_v\, dV - \iint_S k\nabla T \cdot n\, dS$$

$$+ \iiint_V f_v \cdot u\, dV + \iiint_V P\nabla u\, dV +$$

$$+ \iint_S (\overline{\overline{\tau}}^*\, n)\cdot u\, dS$$

Some order of magnitude analysis

A deeper examination of the momentum and energy equations will allow us to know which terms are dominant and which ones could be neglected. This analysis is very useful and must be done prior any kind of CFD simulation in order to have a wider knowledge of the dominant physics of our problem and act accordingly during the simulation setup.

Prior to the analysis, we must express our equations in terms of volume integrals in order to easily compare them. We will also assume that the source term is the gravitational force. In this case it is very usual to use a redefined pressure given by the expression:

$$P' = P - P_h = P + \rho_\infty g z$$

With z, the spatial component in the direction of gravity.

$$\underbrace{\frac{\partial}{\partial t} \int\int\int_V \rho u \, dV}_{(0)} + \underbrace{\int\int\int_V (\nabla \cdot \rho u) u \, dV}_{(1)}$$

$$= \underbrace{\int\int\int_V \nabla P' \, dV}_{(2)} + \underbrace{\int\int\int_V \nabla \cdot \overline{\overline{\tau}} \, dV}_{(3)}$$

$$+ \underbrace{\int\int\int_V (\rho - \rho_\infty) g \, dV}_{(4)}$$

We now make the assumption that the control volumes are small enough (and equal to V) to consider that the fluid properties do not vary inside them; this approach will also be followed when we discretize the equations in section 0. It is also necessary to know a priori some characteristics values of our flow, for the momentum equation we need: a characteristic length of the flow L_c (could be the length of our car or the chord of an airfoil), a characteristic velocity u_∞ (in external flow, the outfield velocity) and a characteristic frequency of the oscillations of fluid properties f_c (this is usually not easy to get but can be known by previous experiences in CFD or WT).

We will estimate the time and space derivatives as:

$$\frac{\partial}{\partial t} \sim \frac{1}{T_c} = f_c$$

$$\nabla \sim \frac{1}{L_c}$$

Now we are ready to estimate the different terms of the momentum equation:

$$(0) \sim f_c \rho u_\infty V$$

$$(1) \sim \frac{1}{L_c} \rho u_\infty^2 V$$

$$(2) \sim \frac{1}{L_c} \Delta P V$$

$$(3) \sim \frac{1}{L_c} \mu \frac{1}{L_c} u_\infty V = \frac{1}{L_c^2} \mu u_\infty V$$

$$(4) \sim \Delta \rho g V \quad (for\ bouyancy\ force)$$

If we know compare the terms with the convective term (1):

$$\frac{(0)}{(1)} \sim \frac{L_c f_c \rho u_\infty}{\rho u_\infty^2} = \frac{f_c L_c}{u_\infty} = St$$

$$\frac{(2)}{(1)} \sim \frac{\Delta P}{\rho u_\infty^2} \sim \mathcal{O}(1) \Rightarrow \Delta P \sim \rho u_\infty^2$$

$$\frac{(3)}{(1)} \sim \frac{\frac{1}{L_c^2} \mu u_m}{\frac{1}{L_c} \rho u_\infty^2} = \frac{\mu}{\rho u_\infty L_c} = \frac{\nu}{u_\infty L_c} = \frac{1}{Re}$$

$$\frac{(4)}{(1)} \sim \frac{\Delta \rho g}{\frac{1}{L_c} \rho u_\infty^2} = \frac{Gr}{Re^2}$$

It has been exposed that an analysis of the order of magnitude of the different terms can give the engineer an idea of the relative weight of each term in his/her simulation. This can help us neglect terms as the gravitational force if their order of magnitude are low compared to the convective term.

If we apply the same procedure to the energy equation, firstly putting all the equation in terms of volume integrals and expressing the equation in terms of P' as we did in the momentum equation:

$$\frac{\partial}{\partial t}\int\int\int_V \rho h\, dV + \underbrace{\int\int\int_V \rho u \cdot \nabla h\, dV + \frac{\partial}{\partial t}\int\int\int_V \rho \frac{1}{2}|u|^2 dV}_{(0e)}$$

$$+ \int\int\int_V \rho u \cdot \nabla\left(\frac{1}{2}|u|^2\right) dV$$

$$= \underbrace{\frac{\partial}{\partial t}\int\int\int_V P'\, dV}_{(1e)} + \underbrace{\int\int\int_V u \cdot \nabla P'\, dV}_{(2e)}$$

$$+ \underbrace{\int\int\int_V \rho \dot{q}_v\, dV}_{(3e)} - \underbrace{\int\int\int_V \nabla(k\nabla T)\, dV}_{(4e)}$$

$$+ \underbrace{\int\int\int_V (\rho - \rho_\infty)g \cdot u\, dV}_{(5e)}$$

$$+ \underbrace{\int\int\int_V \nabla \overline{\overline{\tau}}^i \cdot u\, dV}_{(6e)}$$

In order to estimate the different terms of the energy equation, we will need additional information from our flow, a characteristic temperature drop ΔT that can be obtained, for instance, from the difference between the outfield temperature and a hot wall temperature. The orders of magnitude are the following:

$$(0e) \sim \rho u_\infty \frac{\Delta h}{L_c} V = \rho u_\infty \frac{C_p \Delta T}{L_c} V = \rho u_\infty \frac{C_p(T_w - T_\infty)}{L_c} V$$

$$(1e) \sim \frac{L}{u_\infty}\Delta P' V = f_c \Delta P' V$$

$$(2e) \sim u_\infty \frac{1}{L_c}\Delta P' V$$

$$(4e) \sim \frac{1}{L_c}k\frac{1}{L_c}\Delta T\, V = \frac{1}{L_c^2}k(T_w - T_\infty)V$$

$$(5e) \sim \Delta\rho g u_\infty V \ (for\ bouyancy\ force)$$

$$(6e) \sim \frac{1}{L_c} \mu \frac{1}{L_c} u_\infty^2 \, V = \frac{1}{L_c^2} \mu u_\infty^2 V$$

If we know compare the terms with the convective enthalpy transport term $(0e)$:

$$\frac{(1e)}{(0e)} \sim \frac{\dfrac{L}{u_\infty} \Delta P'}{\rho u_\infty \dfrac{C_p (T_w - T_\infty)}{L_c}} = Ec$$

$$\frac{(2e)}{(0e)} \sim \frac{u_\infty \dfrac{1}{L_c} \Delta P'}{\rho u_\infty \dfrac{C_p (T_w - T_\infty)}{L_c}} = Ec$$

$$\frac{(4e)}{(0e)} \sim \frac{\dfrac{1}{L_c^2} k (T_w - T_\infty)}{\rho u_\infty \dfrac{C_p (T_w - T_\infty)}{L_c}} = \frac{1}{Pe} = \frac{1}{Pr \, Re}$$

$$\frac{(5e)}{(0e)} \sim \frac{\Delta \rho g u_\infty}{\rho u_\infty \dfrac{C_p (T_w - T_\infty)}{L_c}} = \frac{Ec}{Fr}$$

$$\frac{(6e)}{(0e)} \sim \frac{\dfrac{1}{L_c^2} \mu u_\infty^2}{\rho u_\infty \dfrac{C_p (T_w - T_\infty)}{L_c}} = \frac{Ec}{Re}$$

For example, it is very often to neglect the effect of viscous dissipation when $\frac{(6e)}{(0e)} \ll 1$.

If we found in our previous study that it is very likely that we are in this situation, we can do not consider it in our CFD simulation and save computational resources and time. Moreover, this kind of approach will give us a much clearer image of the predominant effects in our flow and as a consequence, we will have a better understanding of its nature. Again, we have shown the importance of this previous calculation.

Mathematical behavior

Now that we have derived the governing equations, it is interesting analyzing them from a mathematical point of view. This analysis, as the order of magnitude analysis, will give us a wider picture of the behavior of our flow prior to numerically solve it.

In order to do so, we will need the equations in their differential form so firstly we will transform our integral form equations. This step is almost straightforward, we only need to put all terms in our integral equations in terms of volume integral, as we did in section 0. However, it is important to notice that to make this transformation we are applying the Gauss theorem that makes the assumption that our fluid properties are differentiable, so in case of expecting discontinuities in our flow (like shock waves) we must be careful. Making this transformation to our original equations we obtain the following set of equations:

- $$\frac{\partial}{\partial t}\int\int\int_V \rho\, dV + \int\int\int_V \nabla(\rho u)\, dV$$

- $$\frac{\partial}{\partial t}\int\int\int_V \rho u\, dV + \int\int\int_V (\nabla\cdot\rho u)\, u\, dV = \int\int\int_V \nabla P'\, dV + \int\int\int_V \nabla\cdot\overline{\overline{\tau}}\, dV +$$
  $$+\int\int\int_V (\rho - \rho_\infty)g\, dV$$

- $$\frac{\partial}{\partial t}\int\int\int_V \rho h\, dV + \int\int\int_V \rho u\cdot\nabla h\, dV + \frac{\partial}{\partial t}\int\int\int_V \rho\frac{1}{2}|u|^2 dV +$$
  $$\int\int\int_V \rho u\cdot\nabla\left(\frac{1}{2}|u|^2\right) dV = \frac{\partial}{\partial t}\int\int\int_V P'\, dV + \int\int\int_V u\cdot\nabla P'\, dV + \int\int\int_V \rho q_v\, dV -$$
  $$-\int\int\int_V \nabla(k\nabla T)\, dV + \int\int\int_V (\rho - \rho_\infty)g\cdot u\, dV + \int\int\int_V \nabla\overline{\overline{\tau}}\cdot u\, dV$$

If we make our control volume infinitesimal; that set of equations is equivalent with the following:

- $$\frac{\partial}{\partial t}\rho + \nabla(\rho u)$$
- $$\frac{\partial}{\partial t}\rho u + (\nabla \cdot \rho u)\, u = \nabla P + \nabla \cdot \overline{\overline{\tau}} + f_v$$
$$\frac{\partial}{\partial t}\rho h + \rho u \cdot \nabla h + \frac{\partial}{\partial t}\rho \frac{1}{2}|u|^2 + \rho u \cdot \nabla\left(\frac{1}{2}|u|^2\right) = \frac{\partial}{\partial t}P + u \cdot \nabla P + \rho \dot{q}_v - \nabla(k\nabla T)$$
- $$+ f_v \cdot u + \nabla\overline{\overline{\tau}} \cdot u$$

Which is a system of partial differential equations (from now on PDE). Without entering in too much details (see reference [1] for a complete explanation), PDE equations can be classified as:

- Hyperbolic
- Parabolic
- Elliptic

Turbulence phenomena

Introduction

At this point we have already introduced the governing equations and discussed their mathematical behavior. However there is still an important point we have not touched, turbulence. Turbulence is a phenomenon that arises at a certain regime and brings instabilities to our equations. In nature, most of the flows are turbulent and from ancient times it has been observed and discussed although a more CFD-oriented treatment comes from the XX century with Kolmogorov's theory.

Leonardo da Vinci studied and described, with his drawings, this phenomenon, sketching different flow patterns. For example, in the following illustration we can appreciate a water flow coming from a channel and approaching a body of water from above its surface.

Figure 0.6 Da Vinci's illustration of falling water over a body of fluid.

Turbulence arises in fluids flows when the inertial forces are so strong that the viscous dissipation cannot damp them. If we now recall what we introduced in section 0, where we analyze the proportional weight of each term of the equations, we may remember that the ratio between inertial forces and viscous forces was given by Reynolds number:

$$Re = \frac{inertial\ forces}{viscous\ forces} = \frac{\frac{1}{L_c}\rho u_\infty^2}{\frac{1}{L_c^2}\mu u_\infty} = \frac{\rho u_\infty L_c}{\mu} = \frac{u_\infty L_c}{\nu}$$

Some characteristics of these flows are the following:

- Irregularity and randomness: turbulent flows area chaotic and intrinsically unsteady. In them, a great variety of spatial scales, as observed in Figure 0.1 in the different eddy sizes, and time scales can be observed. However, if a temporal average is applied to the flow, steady solutions for this average flow can be sometimes obtained.

• Diffusivity: turbulence boosts transport by diffusion. The chaotic characteristic of the flow improves its mixing and hence the diffusivity. This characteristic enhances the heat transfer in turbulent flows and also delays the separation of the boundary layer due to a greater momentum exchange in the wake of the objects.

• Three-dimensional: turbulent flows are always three-dimensional although two-dimensional solutions can be obtained by averaging the flow as we mentioned before.

• Coherent structures: even in the chaos of turbulent flows, coherent structures [4] can be observed as regions of concentrated vorticity, characteristic and flow-specific organization, recurrence, appreciable lifetime and scale. Some of this structures can are presented in Figure 0.7.

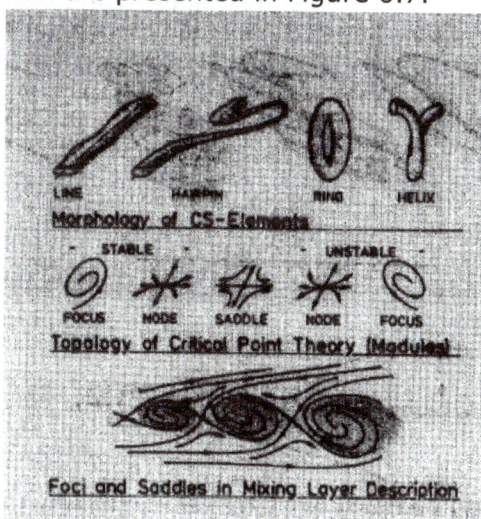

Figure 0.7 Some coherent structures observed in turbulent flows, extracted from [4].

Kolmogorov's energy cascade

It was not till the 1940s when Andréi Kolmogorov observed an important characteristic of turbulent flows. Inside all the chaos he found a pattern. He observed that the small eddies appeared to have a greater vorticity while the big ones contained a great proportion of kinetic energy. From this observation he stated that the energy of the eddies in turbulent flows follows a cascade from the big eddies to the smallest ones where the energy is dissipated by viscous heating. In short, we can summarize the principle contributions of his theory:

- The mean flow transfer energy to the biggest eddies.
- From them the energy is progressively transferred to smaller and smaller eddies.
- Once the so-called Kolmogorov's scale is reached, the energy of the eddies is transformed into heat by viscous dissipation.

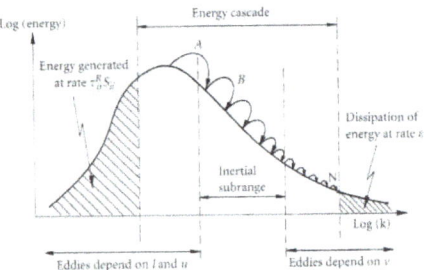

Figure 0.8 Graphical representation of the energy cascade, $k = \dfrac{2\pi}{\lambda}$

Inviability of the direct simulation

In the previous section we have described turbulent flows and their problematic. However, why would we should worry about this annoying instabilities if they completely described by the equations introduced in section 0? Because even with a supercomputer it is very difficult to simulate real-world problems and to make things worse, the vast majority of the flows that are object of study of the motorsport industry are turbulent. A fast calculation of how much time would it takes to perform a direct simulation, also referred as DNS (Direct Numerical Simulation) is presented next:

From partial differential equations to algebraic equations

At this point, we have reviewed the main equations applied to computational fluid dynamics. These equations all together form a system of partial differential equations and are not easy (if even possible) to solve in complex domains as the ones employed in motorsport applications. The approach followed to tackle this issue, in a *divide and conquer* fashion, is to convert the continuum domain of our problem in a discrete domain formed by a finite number of elements. This procedure is called *spatial discretization*.

There is another important component in our equations: time. As space, time is also continuum and we will also need to divide it in finite segments in order to convert our differential equations into an algebraic system of equations and to solve it. This procedure is named *time discretization*.

Introduction to Spatial Discretization

First of all we need to split our domain geometrically in smaller subdomains that as a whole conform what we call *mesh*. Before studying how this transformation, from a continuum to a mesh, affects our equations, it is important to study more deeply what is a mesh, the different types of meshes and what we should look for in order to achieve a good mesh.

Mesh Terminology

If the reader has not been in contact with the world of meshes before, it will be useful to illustrate the terminology that we will use in order to make him/her easier-to-follow the next sections. A mesh has the following elements:

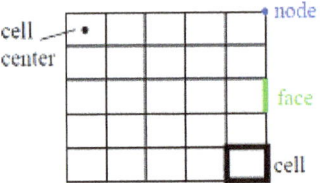

2D computational grid

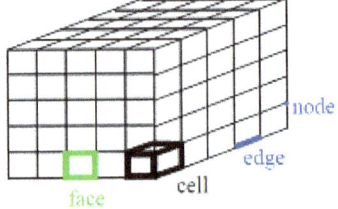

3D computational grid

- Cells: the different volumes in which we split our domain.
- Nodes = mesh points located in the vertex of the cell polygons.

- Cell centers/centroids = centroid of the volume represented by a cell. Variables are usually stored in them in the most used CFD software.
- Edges
- Faces
- Zones = a group of cells, faces or nodes.
- Domain = group of nodes, faces and cell zones.

Mesh types

We classify meshes based on different criteria:

Based on the number of spatial dimensions:

- 1D
- 2D

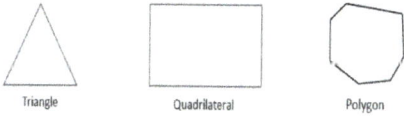

Triangle   Quadrilateral   Polygon

Figure 0.9 2D Mesh element types

- 3D

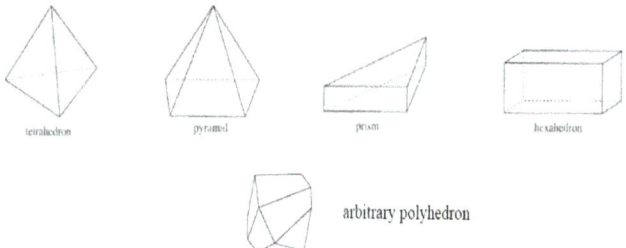

tetrahedron   pyramid   prism   hexahedron

arbitrary polyhedron

Figure 0.10 3D Mesh Element Types

The triangular elements (2D) or the tetrahedrons, also referred as tris and tetras, are employed when meshing complex geometries due to their great adaptability. The quadrilaterals (2D) and hexahedrons (3D), also noted as quads and hexas, allow us to save number of elements and are usually employed in structured meshes (that will be explained later on). The remaining polyhedron elements are usually employed as transitional elements between tetras and hexas and in more complex meshing techniques as the QUIMERA technique [4].

The concept of structured mesh is usually misused and can lead to confusion. Sometimes, people refer a mesh as structured just when it is composed by quads or hexas. What really distinguishes a structured mesh from an unstructured one is that in structured meshes a biunivocal correlation (two-way relationship) between two or three Cartesian indexes (depending if it is 2D or 3D) and each cell must exist. Hence, it is necessary the existence of a function that transforms the Cartesian indexes in the corresponding cell. This characteristic is very handy when allocating our mesh information in a computer and makes this meshes more efficient in memory basics.

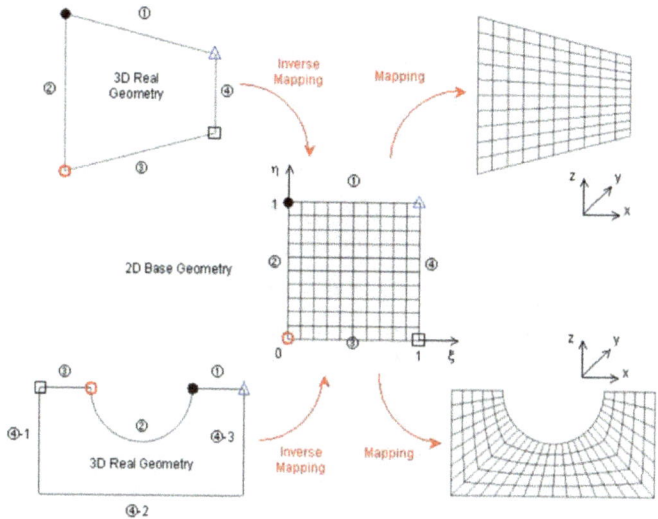

Figure 0.11 Example of structured mesh and the geometrical transformations required

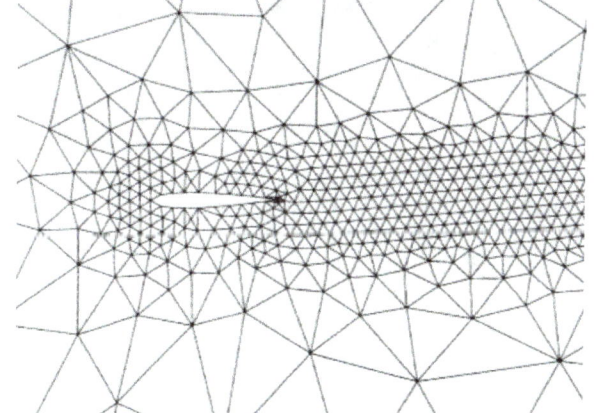

Figure 0.12 Example of unstructured mesh

Based on element connectivity: Conformal/non-conformal

Another way of classifying meshes I based on whether they are conformal or not. A mesh is conformal when each face from a cell is in contact with a unique cell of a neighbor cell. This is better understood through the following examples:

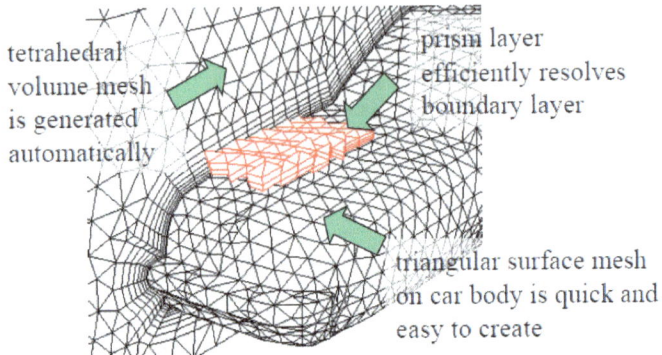

tetrahedral volume mesh is generated automatically

prism layer efficiently resolves boundary layer

triangular surface mesh on car body is quick and easy to create

Figure 0.13 Example of conformal mesh

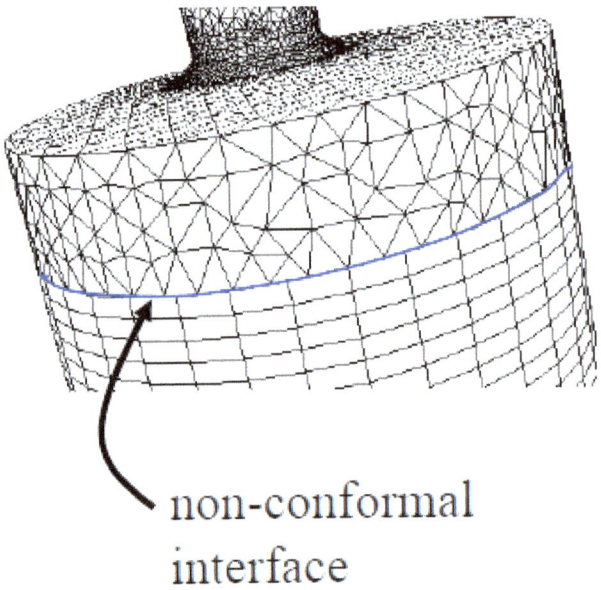

non-conformal interface

Figure 0.14 Example of non-conformal mesh

As can be observed from the previous Figures, non-conformal meshes implies the existence of the so-called non-conformal interfaces which may be treated carefully and can origin numerical errors. Generally, unless is not possible in any other way, it is advisable to employ conformal meshes.

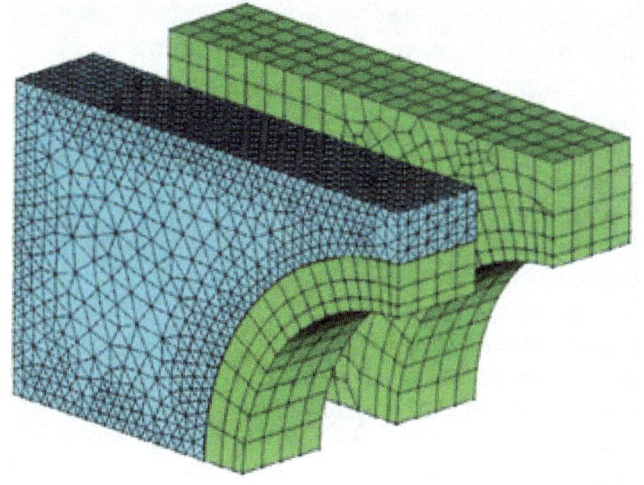

Meshing goals

When meshing, we will try to achieve the following goals (some of them are linked so a compromise may be taken):

•   Get closest approximation of the real-world geometry, especially in those areas that may be critical for our problem.

o   Example: Even in Motorsport industry where usually great computational resources are available, it is very uncommon to mesh small elements as could be bolts, rivets or other minor details that do not influence the flow resolution. Avoiding these details does not penalize the accuracy of the solution and it could even improve it as more resources will be available for meshing more important zones.

- High mesh resolution in those zones where great gradients of the fluid variables are expected. Example: boundary layers.
- Keeping the cell count as low as possible.
- Correct quality parameters (will be discussed in more detail in next section).
  o Example: A usual parameter employed is skewness. At a practical level is advisable to have skewness under 0.9 for the volumetric mesh and under 0.8 for the superficial mesh. Another critical parameter is the non-orthogonality that should be kept under 65º.

Quality parameters

There are a great variety of quality parameters for determining the condition of a mesh. We will only name here a few that are probably the widest used.

Skewness

Measures the deviation of a cell from the ideal cell. The scale employed depends on the software but it is usual to consider 0 as the perfect cell and 1 as the worst possible. For example, for a triangle:

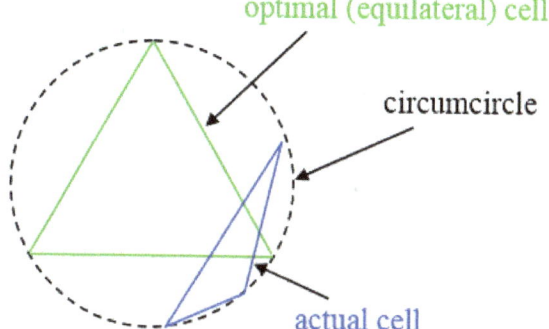

optimal (equilateral) cell

circumcircle

actual cell

Figure 0.15 Skewness measure

$$Skewness = \frac{optimal\ cell\ size - cell\ size}{optimal\ cell\ size}$$

Note that the exact formula behind the parameter can vary but the philosophy behind it is untouched. Skewness gives us an idea of where we have sharp elements that can harm the accuracy of our solution.

Another usual expression applied for quads and the generalization of it are the following:

$$Skewness(quad) = \max\left[\frac{\theta_{max} - 90}{90}, \frac{90 - \theta_{min}}{90}\right]$$

$$Skewness(general) = \max\left[\frac{\theta_{max} - 90}{180 - \theta_e}, \frac{90 - \theta_{min}}{\theta_e}\right]$$

Aspect Ratio

Represents the ratio between the biggest magnitude of an element and the smallest one. An aspect ratio of 1 is a perfect cell and the greatest the aspect ratio the more worried we may be about it. However, in order to keep number of elements low, it is usual to have high aspect ratio in the boundary layers. This is not a big issue as the greatest gradients of the variables are concentrated in the smallest direction of our cells.

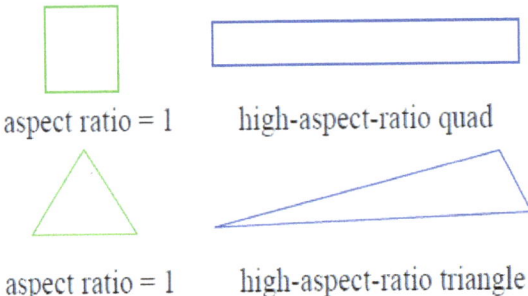

Figure 0.16 Aspect ratio examples

Smoothness

The smoothness gives us indication of how fast our cells change their size, being advisable to keep a no greater than 1.25-1.3 ratio between neighbor cells. Of course, as any other indication given in this section, it highly depends on the nature of the flow in the area we are analyzing so must be taken carefully.

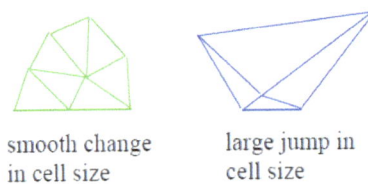

Figure 0.17 Examples of different smoothness meshes

Non-orthogonality

The non-orthogonality parameter represents the deviation between the vector joining the cell centroids and the vector normal to the interface and is measured in degrees.

This can be easily appreciated in the following picture:

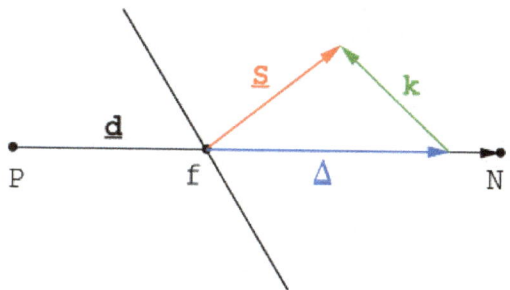

Figure 0.18 Illustration of the different vectors involved in the measure of non-orthogonality

As I have previously stated, there a huge amount of different quality parameters that can help us determine where our meshes can improved. There is no such a master parameter; however, the employment of skewness is widely accepted in the CFD community. Another important point is to keep a methodology between simulations, keeping the quality parameters in a predetermined range. In this way, results will be more comparable and representative.

FVM: the most employed spatial discretization method for CFD

There are several spatial discretization methods. Probably the most important are the finite difference method (FDM), the finite element method (FEM) and finite volume (FVM). We will outline the main different between this approaches.

The *finite difference method* is the oldest of all and approximates the differential terms in the equations by using truncated Taylor series. The basic FDM needs a structured grid and it is difficult to employ it with complex geometries, although improvements have been made lately. It employs the differential form of the equations and the variable values are stored in the nodes of the mesh. Its use to solve fluid flow problems is somewhat limited to easy geometries but it is useful from an academic point of view as it is probably the easiest spatial discretization method to understand. The reader could therefore find a lot of online resources employing this method so we will not get deeper.

The *finite element method* is based on a completely different approach. In order to transform the partial differential equations into algebraic equations, it uses a particular form of the equations, the weak equation where the different terms of the equations are multiplied by a function called test function. This is a rough summary and the reader is encouraged to learn more about this method in [5] if specially interested. Another characteristic is that finite element methods impose a variation of variables through elements being probably the best method for coarse meshes. On the other hand, the computational effort is also high. They are very popular in mechanical and crash simulations and although there are commercial software (Comsol) and open-source (Elmer) that employ them to solve fluid flow, they are outpaced by finite volume methods in this purpose.

Finally, we have the *finite volume methods* which are widely employed in the CFD world, having a great variety of commercial (Fluent, StarCCM+) and open-source software (OpenFOAM). One of the greatest qualities of this method is that the use of the conservative integral form of the governing equations is applied to each cell and it is considered as the most successful in guarantying the conservation of mass, momentum and energy at each cell.

The values of the variables are usually stored at the cell centroids (there are also codes in which they are store at the centre of faces but are more uncommon). Another characteristic is that the shape of the volumes employed to discretize the domain can be arbitrary, leading to greater alternatives during the meshing procedure.

Also when solving compressible flows where discontinuities can arise, the FVM has been shown as the most robust.

Face values schemes

As we have stated earlier, in FVM methods the values of the different variables are allocated in the cell centroids. When during the calculation process the values of the variables (or their derivatives) are required in the faces of the cells, which is indeed very typical, assumption about the variation of the variables between cells are needed.

We will now discuss some of the most employed schemes that calculates this face values using the values of the surrounding cells. This schemes are critical in order to achieve a good stability and accuracy in our runs and we should look for schemes that satisfy:

- Conservativeness: our schemes must ensure the conservation of the fluid property.
- Boundedness: face values must be in the range of the neighbor cells when solving linear problems.
- Transportivenees: our schemes must address the effect of convection and its strength in opposition with diffusion.

First order upwind

This is one of the simplest schemes and widely used as a first approach to our solution. Assumes the face value as the value of the cell in the upstream direction of the flow.

This approach makes this scheme very diffusive which enhances the stability of our run but in contrast harms its accuracy.

A good strategy is start using this scheme and move on to a higher order scheme later in the calculation process, especially in flows highly dominated by the convective flux.

This scheme can lead to erroneous solutions, especially when the grid is not aligned with the main flow direction. This phenomena is usually referred as *false diffusion* and we can see an example in the following figure:

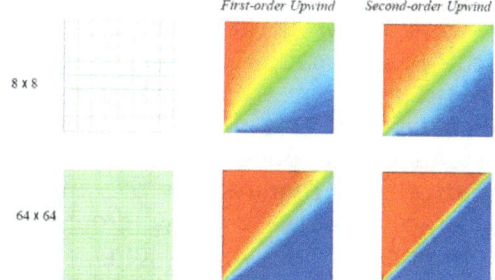

First-order Upwind   Second-order Upwind

8 x 8

64 x 64

Figure 0.19 Example false diffusion problems when using firs order upwind schemes in unaligned meshes

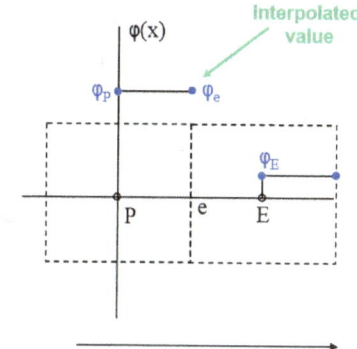

$\varphi(x)$

Interpolated value

φ_P φ_e

φ_E

P   e   E

Flow direction

Figure 0.20 Illustration of the first order upwind scheme

Linear/Central differencing scheme

Another simple scheme is the well-known linear scheme, which is the one we will use for simplicity in our Matlab examples. It assumes the face value as a linear interpolation between the neighbor cells. At this point the reader can easily understand why a good mesh, without extremely distorted cells, is required, as the interpolation for finding face values usually employ geometric characteristics as the vector joining two adjacent cells and when these cells are distorted interpolation errors easily arise. Its accuracy is greater than the one from the upwind scheme but it can produce oscillations (leading even to unbounded solutions) under some flow characteristics (Péclet number, which relates convective and diffusive transport, larger than 2).

A big downside of this scheme is that it does not assess the influence of the convection for high Pe number affecting dramatically the transportiveness of the scheme.

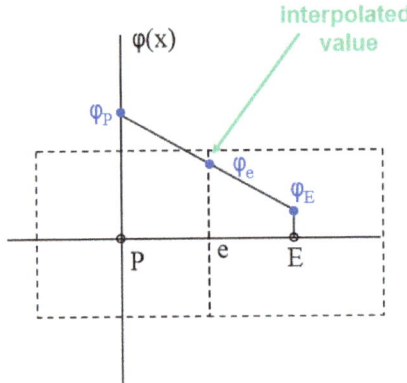

Figure 0.21 Illustration of the linear scheme
Power-law scheme

The big downside of the central scheme is solved with the power-law scheme. In this scheme the face values are calculated from the exact solution of a 1 dimensional convection-diffusion problem, which is formulated as:

$$\frac{\partial}{\partial x}(\rho u \phi) = \frac{\partial}{\partial x} D \frac{\partial \phi}{\partial x}$$

Considering ρu and D as constants in our domain, we can obtain the exact solution:

$$\frac{\phi(x) - \phi_P}{\phi_P - \phi_E} = \frac{\exp\left(Pe\frac{x}{L}\right) - 1}{exp(Pe) - 1}$$

Where $Pe = \frac{\rho u L}{D}$

Leading to the following solutions depending of the Pe number: From the figure is observed that at high Péclet number the scheme tends to an upwind scheme.

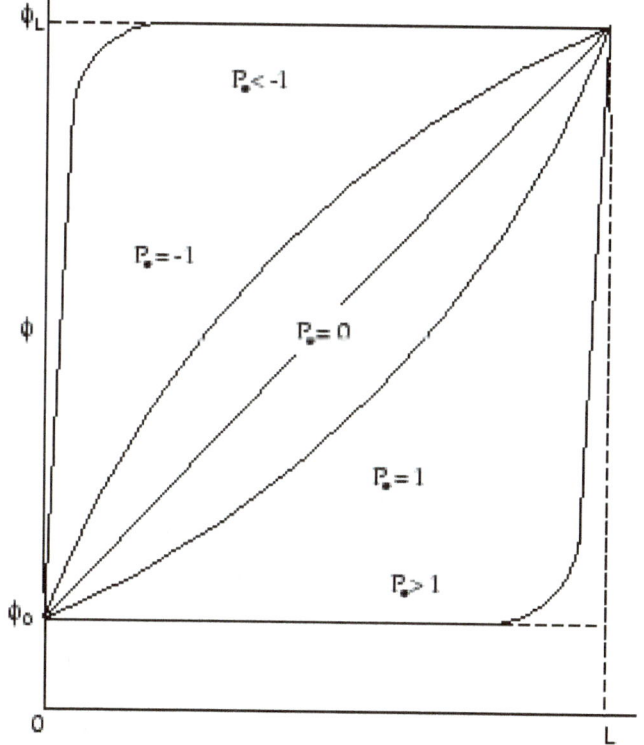

Figure 0.22 Solutions of the power-law scheme

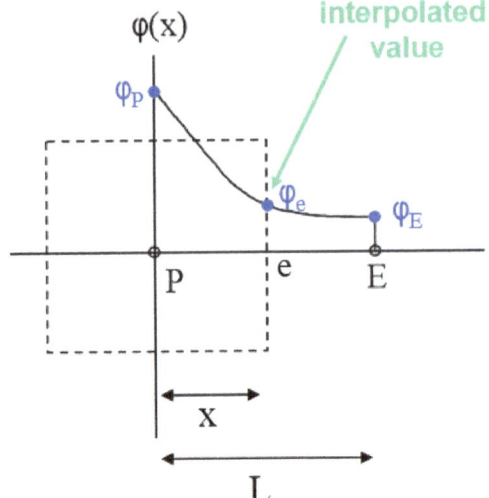

Figure 0.23 Illustration of the power-law scheme

Second-order upwind

The second-order upwind scheme is an enhancement to the first-order upwind that tries to solve the accuracy problems of the first-order scheme. To do so, this scheme uses values not only from the first upstreaming cell but also the second one.

This can lead to an important issue, when this scheme is applied in areas of strong varying gradients, it can lead to a face value out of the range of the two neighbor cells of the face which conduct us to erroneous solutions. Therefore it is necessary to set limiter when using this kind of scheme, guaranteeing that the face value is in the proper range.

The proper implementation can vary in the different commercial software. Here, the implementation from the Fluent User Guide is presented:

$$\phi_e = \phi_P + (\nabla\phi)_E \cdot r$$

Being r, the vector connecting the centroid E and the face center of e.

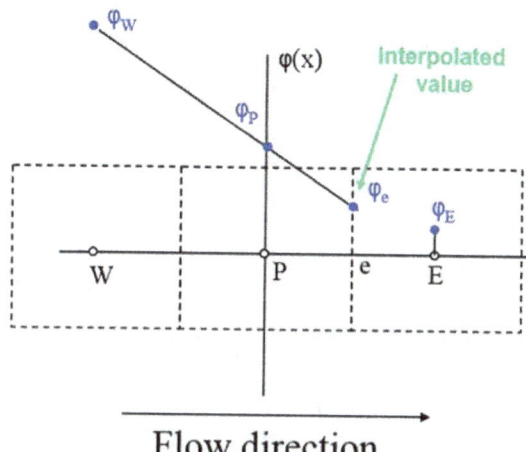

Figure 0.24 Illustration of the second-order upwind scheme

QUICK scheme

This scheme, which name stands for Quadratic Upwind Interpolation for Convective Kinetics, has almost the same upsides and downside as the second-order upwind. On the one hand, the accuracy of this scheme is better than first order upwind or even second-order (not always) but on the other hand, it can also lead to overshoots or undershoots in the face values and has to be applied in conjunction with limiters.

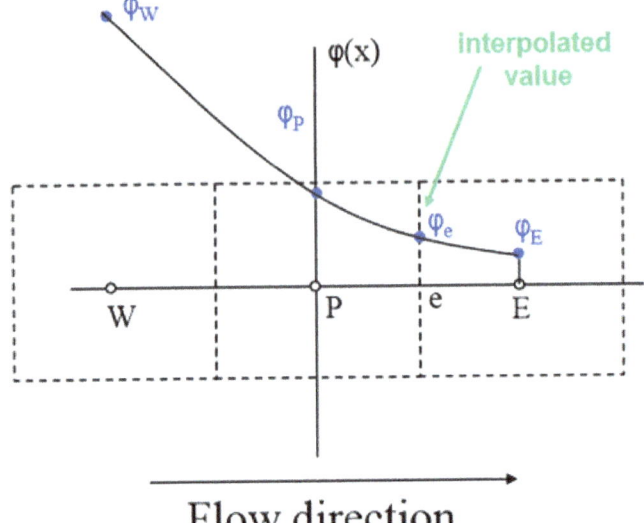

Flow direction

Figure 0.25 Illustration of the QUICK scheme
The correspondent accuracy order of the schemes here presented can be obtained by a Taylor series expansion and it is easily obtainable and can be consulted directly in [6].

Gradient reconstruction methods

During the discretization of the governing equations employing the FVM method is also necessary to convert the different gradients and derivatives in our equations in terms of cell centroids values. This can be accomplished employing the so-called Green-Gauss theorem which can be written in a discrete form as:

$$(\nabla \phi)_c = \frac{1}{Volume_{cell}} \sum_{faces} \phi_{face} \mathbf{n}$$

The face values involved in the expression are not computes using the usual face schemes explained in previous section and the way of determining this face values distinguish three fundamental gradient reconstruction methods:

Green-Gauss Cell-Based

The face values are computed as the average of the neighbouring cells ($c0$ and $c1$) values:

$$\phi_{face} = \frac{\phi_{c0} + \phi_{c1}}{2}$$

Green-Gauss Node-Based

In this method the face values are computed as the arithmetic average of the nodal values on the face:

$$\phi_{face} = \frac{1}{N_{nodes}} \sum_{n}^{N_{nodes}} \phi_n$$

Where the node values are computed by a weighted average form surrounding cell neighbours by solving a constrained minimization problem as stated in [7] and [8], which are the references proposed by the Fluent implementation. This scheme is more accurate than the cell-based, especially in tri and tetra meshes.

Least Squares Cell-based

Another scheme is the known as Least Squares. In this scheme, the solution is assumed to vary linearly so we can express the increment of the fluid property between c_0 and c_i as:

$$\Delta\phi = \phi_{ci} - \phi_{c0} = (\nabla\phi)_{c0} \cdot \Delta r_i$$

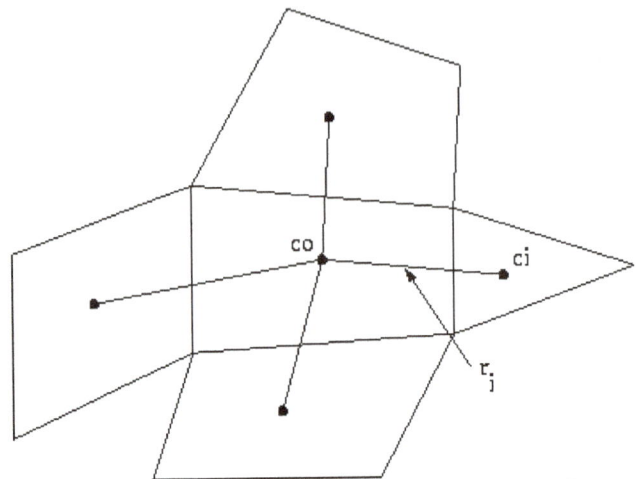

Figure 0.26 Schematic of the used nomenclature, extracted from Fluent User Guide

If we apply this procedure to any surrounding cell, we obtain an overdetermined system of equations that can be expressed in a compact form as:

$$[J](\nabla\phi)_{c0} = \Delta\phi$$

Solving this overdetermined system by the least square technique leads us to the gradient value.

This method is especially recommended for polyhedral meshes. Where it is not usually available the node-based approach. For the rest of meshes, if an accurate solution is needed is recommended the use of the node-based method as is known as more stable.

Example: 2D steady heat transfer

To illustrate all the concepts introduced in this section, we will develop a detailed example involving the solution of a 2D linear and steady heat transfer problem. This will let us fill any gap in the discretization of the equations and give the reader a complete picture of what is going on behind a CFD software.

The problem we will face is a 2D conduction problem where two of the boundaries have a fixed temperature (Dirichlet condition) and the other two have a specific flux (von Neumann condition), in our problem this faces will be isolated so the heat flux will be 0. A schematic of the problem can be observed in the following figure:

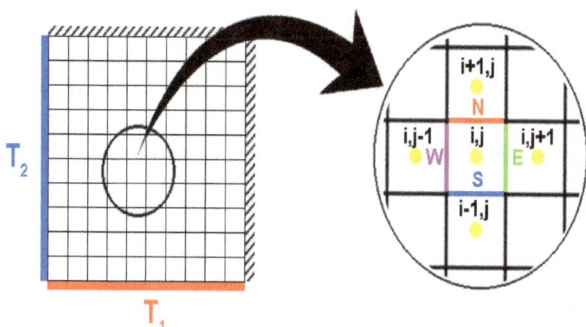

Figure 0.27 Schematic of the 2D conduction example

The governing equation of this problem, in which we assume no internal heat generation, is the following:

$$\nabla \cdot q = 0$$

Which states that the net heat flux across our control volume is 0. If we now apply the Fourier's constitutive law:

$$\nabla \cdot (k\nabla T) = 0$$

Now, in order to apply a finite volume method, we need to obtain the integral form of the governing equation. So we integrate the equation over our control volume:

$$\iiint_V \nabla \cdot (k\nabla T)dV = 0$$

Applying the Gauss theorem:

$$\iiint_V \nabla \cdot (k\nabla T)\, dV = \iint_S k\nabla T \cdot \boldsymbol{n}\, dS = 0$$

At this point, for the sake of simplicity we will assume a constant conductivity all over our domain therefore it can be neglected from the equation, resulting:

$$\iint_S \nabla T \cdot \boldsymbol{n}\, dS = 0$$

If we now discretize this integral assuming constant values of the fluid variables in the faces:

$$\sum_{face\ i}^{nFaces} (\nabla T \cdot \boldsymbol{n})_{face\ i}\, S_{face\ i} = 0$$

Even in a 2D simulation like the one we are here approaching, in FVM we decompose the domain in cells. Imagine the domain is decomposed in cells with a unitary z width so when we speak about volumes we will be referring to surfaces and when we need the surface of the face it will actually be the length. In the same way, the effect of the back and top faces of each cells (which are parallel to the XY plane) will be ignored. In our simple mesh we can write it like:

$$(\nabla T_E)_x S_E - (\nabla T_W)_x S_W + (\nabla T_N)_y S_N - (\nabla T_S)_y S_S = 0$$

As our mesh is going to be completely uniform, all the face surfaces are going to be equal so:

$$\boxed{(\nabla T_E)_x - (\nabla T_W)_x + (\nabla T_N)_y - (\nabla T_S)_y = 0}$$

This last is the equation that must be satisfied in every cell of our domain, giving us a equation for each of the cells which leads to determined system of equations. Now, we will carefully explain how to express the equation in terms of our variables that are the centroid Temperatures.

First, we need to compute the gradients. For this purpose we will employ the Green-Gauss cell-based approach explained in previous sections:

$$(\boldsymbol{\nabla T})_{i,j} = \frac{S}{V}(T_{i,j,E} - T_{i,j,W}, \qquad T_{i,j,N} - T_{i,j,S})$$

We now need to make an assumption about the variation of the temperature in order to obtain the face values, we will choose a linear scheme and as our grid is equally spaced:

$$T_{i,j,N} = \frac{1}{2}(T_{i,j} + T_{i+1,j}); \qquad T_{i,j,S} = \frac{1}{2}(T_{i,j} + T_{i-1,j});$$

$$T_{i,j,E} = \frac{1}{2}(T_{i,j} + T_{i,j+1}); \qquad T_{i,j,W} = \frac{1}{2}(T_{i,j} + T_{i,j-1});$$

So, our original expression for the gradient:

$$(\nabla T)_{i,j} = \frac{S}{2V}(T_{i,j} + T_{i,j+1} - T_{i,j} - T_{i,j-1},$$

$$T_{i,j} + T_{i+1,j} - T_{i,j} - T_{i-1,j}) =$$

$$= \frac{S}{2V}(T_{i,j+1} - T_{i,j-1}, \qquad T_{i+1,j} - T_{i-1,j})$$

Now we have an expression that gives us the value of the temperature gradient at a cell centroid. However in our equation we need this value at the faces so again we need to apply our face scheme to the gradient so:

$$\nabla T_{i,j,N} = \frac{1}{2}(\nabla T_{i,j} + \nabla T_{i+1,j}); \qquad \nabla T_{i,j,S}$$

$$= \frac{1}{2}(\nabla T_{i,j} + \nabla T_{i-1,j});$$

$$\nabla T_{i,j,E} = \frac{1}{2}\left(\nabla T_{i,j} + \nabla T_{i,j+1}\right); \qquad \nabla T_{i,j,W}$$
$$= \frac{1}{2}\left(\nabla T_{i,j} + \nabla T_{i,j-1}\right);$$

If we developed the expression and substitute it in the main equation, we would find that for each cell we need values of the surrounding cells and even at a distance of 2 cells in each direction. So we need information in (i,j), $(i+1,j)$, $(i+2,j)$, $(i-1,j)$, $(i-2,j)$, $(i,j+1)$, $(i,j+2)$, $(i,j-1)$ and $(i,j-2)$. The problem comes when we are close to boundaries and we do not have neighbor cells in some directions, there our boundary conditions come into play.

First, we will consider how to calculate the gradient in the cells close to a Dirichlet condition, like the one showed in the following figure:

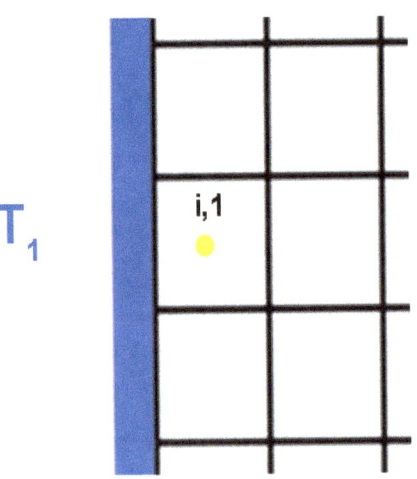

Figure 0.28 Representation of a boundary cell close to a Dirichlet boundary

If we would like to calculate the gradient in this cell, we would need to know the value at the West face. Till now, we have calculated this value as the average of the neighbor cell and the cell itself, here we do not have a neighbor cell but we know the face value as it is our boundary condition. Therefore we can express the gradient in this cells as:

$$\left(\nabla T_{i,1}\right)_x = \frac{S}{V}\left(T_{i,j,E} - T_{i,j,W}\right) = \frac{S}{2V}\left(T_{i,j} + T_{i,j+1} - 2\cdot T_1\right)$$

Another problem would be trying to calculate the gradient not at the last cell but at the proper boundary face. Here and as a result f our governing equation, we will impose that the horizontal component of the heat flux must be constant through the boundary. Therefore we can calculate directly the gradient at the face as being the same as the gradient at the cell centroid:

$$\left(\nabla T_{i,1,W}\right)_x = \left(\nabla T_{i,1}\right)_x$$

Now let us turn our focus into the neighbor cells of a von Neumann condition as the one showed in the figure:

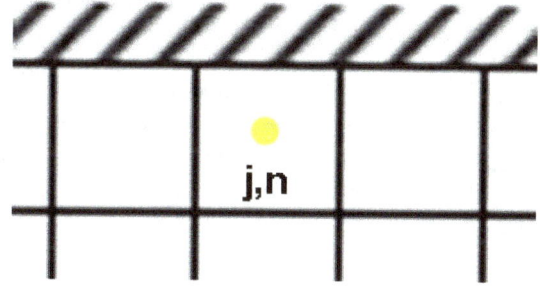

Figure 0.29 Representation of a boundary cell close to a von Neumann boundary

As with the Dirichlet conditions, we have problems using the general expression deduced in first terms as we do not have enough neighbor cells. Here we will use the concept of "ghost cells", these are cells beyond the boundary that we will use to calculate the necessary magnitudes. As we have a 0 flux condition we will suppose that this ghost cell will have the same flux as it must be continuous, therefore the temperature of the first cell and the "ghost cell" is the same to satisfy this condition:

$$\left(\boldsymbol{\nabla}T_{j,n}\right)_y = 0 = \left(\boldsymbol{\nabla}T_{j,n+1=ghost}\right)_y \Rightarrow T_{j,n,N} = T_{j,n+1=ghost}$$
$$= T_{j,n}$$

Moreover, when the gradient of the boundary face is require, it is obviously 0 becouse of the boundary condition.

At this point we have all the "weapons" necessary to tackle our problem so we will start coding it in Matlab. I hope that any doubt that may have arisen will vanish once we finished with the Matlab program. As it was previously mentioned the program that we are going to explain is far from being the most general or most efficient program for this purpose, its main aim is clarify the concepts previously presented.

First we will start defining our different mesh parameters and boundary conditions values. Then we initialize a matrix that will allocate the temperature in two ways for making easier, one in form of a 2D matrix which will help us to visualize the final temperature field and another as a 1D vector which is the way that we will handle temperature during the calculation. The number of unknowns are $m \cdot n$. Remember the classical formulation for solving a linear system of equations:

$$A_{(m \cdot n)x(m \cdot n)} X_{(m \cdot n)x1} = b_{(m \cdot n)x1}$$

The same problem using our nomenclature will be the following:

$$M_{(m \cdot n) x (m \cdot n)} Tvector_{(m \cdot n) x 1} = b_{(m \cdot n) x 1}$$

```
% Problem definition
m = 100; % Number of elements in the x direction
n = 100; % Number of elements in the y direction
T1 = 273.15; % Temperature of the west wall
T2 = 373.15; % Temperature of the south wall
S = 1; % Cell face surface
V = 1; % Cell Volume

% Temperature matrix initialization and
transformation into vector

T = (T1+T2)*ones(m,n);
Tvector = reshape(T, [m*n 1]);
```

We will now start with the matrix assembly stage:

```
disp('Mattrix Assembly');
tic
% Solution matrix assembly
%M = spalloc(m*n, m*n, 4*m*n); % Allow 4
elements per row to be non-zero. This is a more
efficient but less clear to handle sparse matrixes.
M = zeros(m*n);
b = zeros(m*n,1); % independent term vector


for i = 1 : size(T, 1)
   for j = 1 : size(T, 2)
      row = sub2ind(size(T), i, j);

      % North Face

         M(row,:) = M(row,:) + 1/2 *
(calcGrad(i,j,M,T,S,V,2) +
calcGrad(i+1,j,M,T,S,V,2));
         b(row) = b(row) + 1/2 *
(calcIndependentTerm(i, j, S, V, T1, T2) +
```

```matlab
calcIndependentTerm(i+1, j, S, V, T1, T2));

    % South Face

        M(row,:) = M(row,:) - 1/2 *
(calcGrad(i,j,M,T,S,V,2) + calcGrad(i-
1,j,M,T,S,V,2));
        b(row) = b(row) - 1/2 *
(calcIndependentTerm(i, j, S, V, T1, T2) +
calcIndependentTerm(i-1, j, S, V, T1, T2));

    % East Face
        M(row,:) = M(row,:) + 1/2 *
(calcGrad(i,j,M,T,S,V,1) +
calcGrad(i,j+1,M,T,S,V,1));
        b(row) = b(row) + 1/2 *
(calcIndependentTerm(i, j, S, V, T1, T2) +
calcIndependentTerm(i, j+1, S, V, T1, T2));

    % West Face
        M(row,:) = M(row,:) - 1/2 *
(calcGrad(i,j,M,T,S,V,1) + calcGrad(i,j-
1,M,T,S,V,1));
        b(row) = b(row) - 1/2 *
(calcIndependentTerm(i, j, S, V, T1, T2) +
calcIndependentTerm(i, j-1, S, V, T1, T2));

    end
end
toc
```

In this part of the code there are some functions that could be not known by the user:

- sub2ind(size(T), i, j): each element of each 1, 2 or n-dimensional matrix has a linear index that is an alternative way to access them instead of using a set of n indexes.

This functions return this linear index.

- calcGrad(i,j,M,T,S,V,component): this function will be explained deeply later and returns a row vectors with the correspondent coefficients that multiply our unknowns when the gradient of T in the cell (i,j) is calculated.

- calcIndependentTerm(i, j, S, V, T1, T2): in the same way as the previous function it returns the correspondent coefficients during the calculation of the gradient of the cell (i,j) but when no unknowns are involved.

Now we have our matrix assembled and also our independent term and we are in position to calculate the temperature field. For this process we will use two different approaches: (i) use the Matlab \ operator which analyze the matrix and employs the adequate method to solve it (usually a iterative method) and (ii) perform a reordering technique prior to the calculation that can or is automatically done in some CFD software and reduce the bandwidth of the solution matrix, this method is faster in large matrix. At this point the reader could try to solve the system by calculating the inverse of the matrix. This method is completely discouraged but the reader could try it and compare the times required for one and the other method.

```
p = symrcm(M);
R = M(p,p);
```

```matlab
disp('Solving...');
tic
Tvector = M\b;
T = reshape(Tvector, size(T));
toc
T_b = Tvector;


% Cuthill-Mckee
p = symrcm(M);
disp('Solving with Cuthill-McKee...');
tic
Tvector(p) = M(p,p) \ Tvector(p);
toc
```

Finally we will visualize the temperature field in two different ways:

```matlab
%Visualization

imagesc(flipud(T))
colorbar
waitforbuttonpress
contour(T)
colorbar
```

Which lead to the following representations:

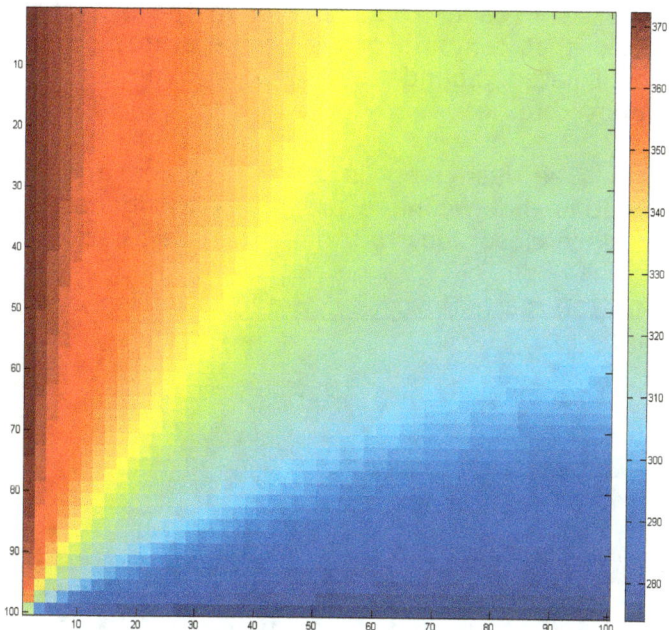

Figure 0.30 Filled contour representation of the temperature field.

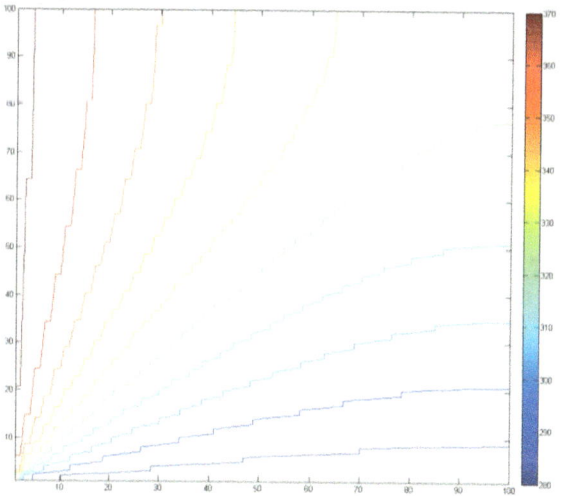

Figure 0.31 Contour representation of the Temperature field.

The reader can observer in the last of the figures that contours of the temperature approach the isolated boundaries perpendicular as it is supposed to be.

The last step is to understand the two functions that we have previously mentioned. Let us take a closer look to the first one calcGrad:

```
function matrixRow = calcGrad(i, j, M, T, S, V, comp)
    %
    % Get the linear index from an element of a 2D matrix
    %
    matrixRow = zeros(1,size(M,1));
    if comp == 1
        if j == 0 || j == 1 % Diritchlet condition left wall
        % Gradient must be continous throught the wall so
        % the gradient of the ghost cell is the same as the first cell
            matrixRow(1, sub2ind(size(T), i, 2)) = S/(2*V);
            matrixRow(1, sub2ind(size(T), i, 1)) = S/(2*V);


        elseif (j == size(T,2)) || (j == (size(T,2)+1)) % Neumann condition isolated
            % Zero gradient
            if j == size(T,2) % The gradient for the ghost cell is just 0
                matrixRow(1, sub2ind(size(T), i, j)) = S/(2*V);
                matrixRow(1, sub2ind(size(T), i, j-1)) = -S/(2*V);
            end
        else
            matrixRow(1, sub2ind(size(T), i, j+1)) =
```

```matlab
S/(2*V);
        matrixRow(1, sub2ind(size(T), i, j-1)) = -
S/(2*V);
    end
  else
    if i == 0 || i == 1 % Diritchlet condition
lower wall
      % Gradient must be continous throught the
wall so
      % the gradient of the ghost cell is the same
as the first cell
        matrixRow(1, sub2ind(size(T), 2, j)) =
S/(2*V);
        matrixRow(1, sub2ind(size(T), 1, j)) =
S/(2*V);


    elseif (i == size(T,1)) || (i ==
(size(T,1)+1)) % Neumann condition isolated
        % Zero gradient
        if i == size(T,1) % The gradient for the
ghost cell is just 0
          matrixRow(1, sub2ind(size(T), i, j)) =
S/(2*V);
          matrixRow(1, sub2ind(size(T), i-1, j))
= -S/(2*V);
        end
    else
      matrixRow(1, sub2ind(size(T), i+1, j)) =
S/(2*V);
      matrixRow(1, sub2ind(size(T), i-1, j)) = -
S/(2*V);
    end
  end


end
```

The first if statement distinguishes whether the user want the x or the y component of the gradient, then it is checked if the cell is close to a boundary, applying in each case the formulae introduced previously in the chapter and it is a good exercise that reader will try to figure them out.

The same approach is employed to the calculation of the coefficients of the independent term caused by each gradient and that are calculated in the following function:

```
function independentTerm = calcIndependentTerm(i, j, S, V, T1, T2)
    %
    % Get the linear index from an element of a 2D matrix
    %

    independentTerm = 0;

        if j == 0 || j == 1 % Diritchlet condition left wall
        % Gradient must be continous throught the wall so
        % the gradient of the ghost cell is the same as the first cell

            independentTerm = independentTerm + T2*S/V;

        end

        if i == 0 || i == 1 % Diritchlet condition lower wall
        % Gradient must be continous throught the wall so
        % the gradient of the ghost cell is the same as the first cell
```

```
        independentTerm = independentTerm +
T1*S/V;

    end



end
```

Time discretization methods

In the vast majority of the CFD techniques, the method of lines is employed.

This means that the spatial variables and the time are discretized separately. In this section we will focus in the part of the CFD solving process that relies on the time discretization. As in the previous section we will support ourselves in the 2D Conduction example to settle the concepts introduced earlier in the section.

First we will introduce a generic partial differential equation over which the different time discretization methods will be studied:

$$\frac{d(\Omega \bar{M} W_I)}{dt} = -\boldsymbol{R}_I$$

Where Ω refers to the volume of our control volume, R is what we will call from here on residual, M is the mass matrix and the sub index I denotes our particular control volume. If we compare this expression with the unsteady heat conduction equation:

$$\frac{d}{dt}(\rho C_p T) = -\nabla \cdot (k \boldsymbol{\nabla} T)$$

If we assume ρ and C_p constant with temperature and therefore with time and considering the diffusivity $\alpha = \frac{k}{\rho C_p}$:

$$\frac{d}{dt}T = -\nabla \cdot (\alpha \nabla T)$$

Which in integral form:

$$\iiint_{VC} \frac{d}{dt}T \, dV = -\iint_S \alpha \nabla T \cdot \boldsymbol{n} \, dS$$

If we discretize it as we have done in previous sections and using the same index nomenclature employed in the general expression:

$$\frac{d}{dt}(T_I \Omega) = -\sum_{face\ i}^{nFaces} (\alpha \nabla T \cdot \boldsymbol{n})_{face\ i} S_{face\ i}$$

It is easily observed that our residual is:

$$\boldsymbol{R}_I = \sum_{face\ i}^{nFaces} (\alpha \nabla T \cdot \boldsymbol{n})_{face\ i} S_{face\ i}$$

And if we would like to solve the steady-state of the equation, this residual must trend to zero leading to the equation solved in the previous section:

$$\boldsymbol{R}_I = \sum_{face\ i}^{nFaces} (\alpha \nabla T \cdot \boldsymbol{n})_{face\ i} S_{face\ i} = 0$$

Returning to the generality that the first equation proportion us, almost all the time discretization methods are contained in the following discretized equation:

$$\frac{(\Omega \bar{M})_I}{\Delta t_I} \Delta W_I^n = -\frac{\beta}{1+\omega} R_I^{n+1} - \frac{1-\beta}{1+\omega} R_I^n$$
$$+ \frac{w}{1+w} \frac{(\Omega M)_I}{\Delta t_I} \Delta W_I^{n-1} \quad ,$$

Where

$$\Delta W_I^n = W_I^{n+1} - W_I^n$$

This expression is second-order accurate in time if:

$$\beta = \omega + \frac{1}{2}$$

Explicit Methods:

Explicit methods are those in which the solution from the next time step is computed explicitly (or directly) from the previous time steps solutions. For example if we make $\beta = 0$ and $w = 0$ in the general expression previously introduced:

$$\frac{(\Omega \bar{M})_I}{\Delta t_I} \Delta W_I^n = -R_I^n$$

If we assume a spatial uniform cell centred grid, this expression gets even simpler:

$$\Delta W_I^n = W_I^{n+1} - W_I^n = -\frac{\Delta t_I}{\Omega_I} R_I^n \Rightarrow W_I^{n+1} = W_I^n - \frac{\Delta t_I}{\Omega_I}$$

This is the simplest explicit method and is known as single-step as the value of the variables for the next time step are calculated in one step from the previous time step data. This simplicity makes these methods so easy to implement.

These methods, however, have some drawbacks. The most important is that the maximum time step is constrained and linked to spatial discretization. So the finer our grid is, the smaller our time step must be in order to keep it stable. When solving fluids this is called the CFL (named after Courant-Friedrichs-Lewy) condition and is stated as follows:

$$\Delta t \leq \frac{\Delta x}{u} C_{max}$$

Where u represents a characteristic velocity of the flow, Δx is a characteristic length of our mesh and C_{max} is the maximum Courant number tolerated. For explicit schemes $C_{max} = 1$. In implicit schemes this number can be substantially higher but even it could be stable is advisable to keep it as low as possible as it also harms accuracy.

Another drawback, which is a particular drawback of the single-stage method, is the instabilities that may arise in this method if a scheme outside first-order-upwind is used. This lead to the development of more sophisticated method as the multistage Runge-Kutta methods. In them the solution update is split in stages (represented by the exponent (i)) in the following way:

$$W_I^{(0)} = W_I^n$$
$$W_I^{(1)} = W_I^{(0)} - \alpha_1 \frac{\Delta t_I}{\Omega_I} R_I^{(0)}$$
$$W_I^{(2)} = W_I^{(0)} - \alpha_2 \frac{\Delta t_I}{\Omega_I} R_I^{(1)}$$
$$\vdots$$
$$W_I^{n+1} = W_I^{(m)} = W_I^{(0)} - \alpha_m \frac{\Delta t_I}{\Omega_I} R_I^{(m-1)}$$

The multistage methods are not perfect though, they require a greater computational effort as the residual must be computed m times in each time step while in the single-stage one computation is enough. When this computation is computer-demanding like in complex fluid flow problems, it is advisable to employ the hybrid explicit methods. These method are essentially the same as the Runge-Kutta ones but do not compute the residual in each stage. More information is available in [2].

Implicit Methods

The implicit methods are more flexible and allow us to employ bigger time steps and are especially suited for solving unsteady flow problems which can lead to a steady-solution. In practice they are also employed in evaluative problems when the size of the problem make unreachable an explicit simulation (because of the tiny time steps required).

Different implicit schemes are obtained just by changing the values of β and ω. A particular case of interest is the so-called 3-point implicit backward-difference scheme, which is recovered by employing $\beta = 1$ and $\omega = 1/2$ and it is 2nd order accurate in time.

It is important to remark that in opposition of what occurred in the explicit schemes, the solution is not computed directly for each cell and it is necessary to solve a system of equations (usually by iterative methods).

The explanation is that when employing implicit schemes we have the term $-\frac{\beta}{1+\omega} R_I^{n+1}$ which is referred to the time step that we are trying to solve so any variable necessary to calculate R_I^{n+1} will appear as an unknown. If we recall the example of the 2D heat conduction, our residual is composed by the net heat flux over the cell. This value is calculated in relation of the surrounding temperatures of the cells so these temperatures will be our unknowns and we will need to solve the resulting system of equations.

For deeper information of this topic the reader is again encouraged to read the corresponding chapter in (CITAR BLAZEK).

Example: 2D unsteady heat transfer

With the following example we will illustrate the simplicity of the one-stage explicit method. We will modify the previously presented code for solving the same example but imposing an initial condition unequal to the steady-state solution. Recalling the single-stage temporal discretization:

$$\Delta W_I^n = W_I^{n+1} - W_I^n = -\frac{\Delta t_I}{\Omega_I} R_I^n \Rightarrow W_I^{n+1} = W_I^n - \frac{\Delta t_I}{\Omega_I}$$

As we have seen previously, for the unsteady heat conduction equation:

$$W_I = T_I$$

$$R_I = \sum_{face\ i}^{nFaces} (\alpha \nabla T \cdot n)_{face\ i} S_{face\ i}$$

Then:

$$T_I^{n+1} = T_I^n - \frac{\Delta t_I}{\Omega_I} \sum_{face\ i}^{nFaces} (\alpha \nabla T \cdot n)_{face\ i} S_{face\ i}$$

Let us now see how to code this:

First we need to modify our previous initialization code to include the diffusivity and the temporal parameters as the time step size and the total time simulated (highlighted in blue):

```
% Problem definition
m = 100; % Number of elements in the x direction
n = 100; % Number of elements in the y direction
T1 = 273.15; % Temperature of the west wall
T2 = 373.15; % Temperature of the south wall
S = 1; % Cell face surface
V = 1; % Cell Volume
diff = 10; % Material diffusivity

deltaTime = 0.1;
totalTime = 1000;
nIter = fix(totalTime/deltaTime) + 1;
```

Now the temperature initialization is critical as it will be our initial condition, in this case we decided to start from the lower temperature and let it warm up. The lines marked in red are no longer needed:

```matlab
% Temperature matrix initialization and transformation into vector

T = (T1)*ones(m,n); % Initial condition
Tvector = reshape(T, [m*n 1]);

disp('Mattrix Assembly');
tic
% Solution matrix assemble
%M = spalloc(m*n, m*n, 4*m*n); % Allow 4 elements per row to be non-zero
M = zeros(m*n);
b = zeros(m*n, 1); % independent term vector
```

Then we need to make an outer loop for each iteration in our main code:

```matlab
for k = 1:nIter
    for i = 1 : size(T, 1)
        for j = 1 : size(T, 2)
            row = sub2ind(size(T), i, j);

%              % North Face
%
%              M(row,:) = M(row,:) + 1/2 * (calcGrad(i,j,M,T,S,V,2) + calcGrad(i+1,j,M,T,S,V,2));
%              b(row) = b(row) + 1/2 * (calcIndependentTerm(i, j, S, V, T1, T2) + calcIndependentTerm(i+1, j, S, V, T1, T2));
%
%              % South Face
%
%              M(row,:) = M(row,:) - 1/2 *
```

```matlab
(calcGrad(i,j,M,T,S,V,2) + calcGrad(i-
1,j,M,T,S,V,2));
%                 b(row) = b(row) - 1/2 *
(calcIndependentTerm(i, j, S, V, T1, T2) +
calcIndependentTerm(i-1, j, S, V, T1, T2));
%
%           % East Face
%               M(row,:) = M(row,:) + 1/2 *
(calcGrad(i,j,M,T,S,V,1) +
calcGrad(i,j+1,M,T,S,V,1));
%               b(row) = b(row) + 1/2 *
(calcIndependentTerm(i, j, S, V, T1, T2) +
calcIndependentTerm(i, j+1, S, V, T1, T2));
%
%           % West Face
%               M(row,:) = M(row,:) - 1/2 *
(calcGrad(i,j,M,T,S,V,1) + calcGrad(i,j-
1,M,T,S,V,1));
%               b(row) = b(row) - 1/2 *
(calcIndependentTerm(i, j, S, V, T1, T2) +
calcIndependentTerm(i, j-1, S, V, T1, T2));
%               if i==1 && j==1
%                   deltaTime*residual(i, j, S, V, T,
T1, T2)
%               end
            Tvector(row) = Tvector(row) +
diff*deltaTime*residual(i, j, S, V, T, T1, T2);
        end
    end
    T = reshape(Tvector, size(T));
    fig = imagesc(flipud(T));
    colorbar
    set(findall(fig,'-
property','FontSize'),'FontSize',18)
    drawnow
end
```

The last five lines in the time loop are intended to graphically represent the temperature field as a movie. Let us now focus on the most important line:

```
Tvector(row) = Tvector(row) +
diff*deltaTime*residual(i, j, S, V, T, T1, T2);
```

If we examine the residual function which is completely new:

```
function res = residual(i, j, S, V, T, T1, T2)
        % North Face

        res =   1/2 * (calcGrad(i,j,T,S,V,2)
+ calcGrad(i+1,j,T,S,V,2)) + 1/2 *
(calcIndependentTerm(i, j, S, V, T1, T2) -
calcIndependentTerm(i+1, j, S, V, T1, T2));


    % South Face

        res = res - 1/2 * (calcGrad(i,j,T,S,V,2)
+ calcGrad(i-1,j,T,S,V,2)) - 1/2 *
(calcIndependentTerm(i, j, S, V, T1, T2) -
calcIndependentTerm(i-1, j, S, V, T1, T2));

    % East Face
        res = res + 1/2 *
(calcGrad(i,j,T,S,V,1) + calcGrad(i,j+1,T,S,V,1))
+ 1/2 * (calcIndependentTerm(i, j, S, V, T1, T2) -
calcIndependentTerm(i, j+1, S, V, T1, T2));

    % West Face
        res = res - 1/2 * (calcGrad(i,j,T,S,V,1)
+ calcGrad(i,j-1,T,S,V,1)) - 1/2 *
(calcIndependentTerm(i, j, S, V, T1, T2) -
calcIndependentTerm(i, j-1, S, V, T1, T2));


end
```

Which is more or less the same we did earlier in our steady script. The main difference is that now the calcGrad function does not return a row in order to assemble it in the solution matrix. Now it returns a numerical value which is the temperature gradient component for that cell based in the previous step temperature field. Let us examine the changes in the calcGrad function:

```
function value = calcGrad(i, j, T, S, V, comp)
    %
    % Get the numerical value of the temperatura gradient
    %
    value = 0;
    if comp == 1
        if j == 0 || j == 1 % Diritchlet condition left wall
            % Gradient must be continous throught the wall so
            % the gradient of the ghost cell is the same as the first cell
            value = value + S/(2*V)*T(i,2);
            value = value + S/(2*V)*T(i,1);


        elseif (j == size(T,2)) || (j == (size(T,2)+1)) % Neumann condition isolated
            % Zero gradient
            if j == size(T,2) % The gradient for the ghost cell is just 0
                value = value + S/(2*V)*T(i,j);
                value = value -S/(2*V)*T(i,j-1);
            end
        else
            value = value + S/(2*V)*T(i,j+1);
            value = value - S/(2*V)*T(i,j-1);
        end
    else
        if i == 0 || i == 1 % Diritchlet condition lower wall
```

```
    % Gradient must be continous throught the
wall so
    % the gradient of the ghost cell is the same
as the first cell
        value = value + S/(2*V)*T(2,j);
        value = value + S/(2*V)*T(1,j);


    elseif (i == size(T,1)) || (i ==
(size(T,1)+1)) % Neumann condition isolated
        % Zero gradient
        if i == size(T,1) % The gradient for the
ghost cell is just 0
            value = value + S/(2*V)*T(i,j);
            value = value - S/(2*V)*T(i-1,j);
        end
    else
        value = value + S/(2*V)*T(i+1,j);
        value = value - S/(2*V)*T(i-1,j);
    end
  end


end
```

And that was all necessary to perform this unsteady simulation. The reader is encouraged to try an implicit or explicit multi-stage modification. The last one should not give him so much extra work.

Non-linearities, how to deal with them

Until this point, when we discretized a differential equation, it led to a linear system of equations. This is far from what it is usually encountered in real-life problems as real-life is far from linear.

Let us examine the steady heat conduction equation again:

$$\iint_S k\boldsymbol{\nabla T} \cdot \boldsymbol{n}\, dS = 0$$

When the conductivity was supposed constant we could neglect its effect. However if we now impose a temperature dependent model, which is closer to the physical phenomena:

$$\iint_S k(T)\boldsymbol{\nabla T} \cdot \boldsymbol{n}\, dS = 0$$

Our discretized equation:

$$\sum_{face\ i}^{nFaces} k(T_{face\ i}) \cdot (\boldsymbol{\nabla T} \cdot \boldsymbol{n})_{face\ i}\ S_{face\ i} = 0$$

Which applied to all cells and with the proper boundary conditions leads to a non-linear system of equations.

There are a huge variety of non-linear solution methods. In this section we will briefly review the most important and easy-to-implement and introduce the more complex and popular among CFD software.

There are some similarities in the methods we are about to introduce and in order to keep a mental structure is important to have them in mind:

- They are iterative methods and introduce a new outer loop (remember that the most efficient way to solve large linear systems is usually an iterative method which will be the inner loop).
- They are strongly dependent on initial iterant even in steady problems. This is a very important point to have in mind also in our simulations with CFD software as the

initialization can strongly affect robustness and convergence.

- A termination criteria is required to stop the loop. A usual approach is to examine the difference between the new and the old iterant. When this difference is small enough we consider the solution as converged.

Fixed point method

From a mathematical point of view, a non-linear system of equations can be expressed as:

$$F(x) = 0, \qquad F : D \subset R^n \to R^n$$

If we review our usual nomenclature to recover F, during our discretization we have obtained a non-linear system of equations like the following:

$$A(x)\, x = b(x) \Rightarrow F(x) = A(x)\, x - b(x)$$

A fixed point iteration is defined by an iteration function Φ:

$$\Phi : U \subset R^n \to R^n$$

We will also need an initial guess or initial iterant:

$$x^{(0)} \in U$$

Our next iterates will be calculated as:

$$x^{(k+1)} = \Phi(x^{(k)})$$

Φ must be continuous and for consistency (and common sense):

$$F(x) = 0 \Leftrightarrow \Phi(x) = x$$

Which means that when we find the exact solution the iterant function must return no change in the unknowns.

There are several ways of transforming F in a fixed point function, we will explain a simple one:

$$F(x) = A(x)\, x - b(x) = 0 \Rightarrow x = A(x)^{-1}b(x) = \Phi(x)$$

Here we have expressed it using the inverse of A, actually as we have repeated several times, is not usual to calculate the matrix inverse. The most common approach is solve the linear system $A(x^k)x^{k+1} = b(x^k)$ employing iterative methods.

The greatest advantage of this method is its simplicity which makes it so easy to implement. On the other hand, its convergence speed is slow and we must be careful to keep x^k in reasonable ranges in order to ensure convergence and to employ a good initial guess.

Newton's method

An improved speed method is the famous Newton's Method, in which we approximate the function $F(x) = 0$ as:

$$F(x) \approx \tilde{F}(x) = F\left(x^{(k)}\right) + J(x^{(k)})(x - x^{(k)})$$

Where J is the Jacobian matrix:

$$J = \left(\frac{\partial F}{\partial x_1} \quad \cdots \quad \frac{\partial F}{x_n}\right)$$

And the iteration can be defined as:

$$x^{(k+1)} = x^k - J\left(x^{(k)}\right)^{-1}F(x^{(k)})$$

Again, although we have expressed the update expression using the inverse of the Jacobian it is not calculated explicitly in most cases.

This method has a fast rate of convergence but has a great drawback, the jacobian must be computed and this process for large problems is computationally expensive. This led to the quasi-Newton methods like the Secant method, which try to approximate the Jacobian employing the Secant and saving a great amount of computational time.

Relaxation on iterative methods

An important concept in CFD is relaxation. This concept, which it is not exclusively employed when solving non-linear systems, is vital in CFD and can determine whether a simulation will or will not converge.

Relaxation consists in reducing the variable change between iterations in order to make less likely that convergence occurs. So when updating the solution:

$$x^{i+1} = x^i + U(x^{i+1(predicted)} - x_i)$$

Where U is called relaxation factor. If U is set to 1, no relaxation is applied. The common procedure is apply a U value under 1, however, in some special situations and to accelerate the solution, values over 1 can be applied.

This technique helps smooth the oscillations that may arise during the simulation, improving the stability. It must be noted that very low values of U can dramatically slow the convergence to the point that we may wrongly think that the case has converged. Experience here plays a key role and help improve convergence in difficult simulations when a great variety of variables are coupled.

Example: non-linear 2D heat transfer

Let us now apply what we have learnt in this section to a real problem. We will take our beloved 2d heat conduction steady problem but now we will impose a temperature-dependent conductivity typical of a stainless steel:

$$k(T) = 14.6 + 1.27e - 2 \cdot T \left[\frac{W}{mK} \right]$$

This will lead to a non-linear problem that we will solve using a fixed-point method. We have introduced two new functions:

- Calc_conductivity(T): which just returns the conductivity for a fixed temperature:

```
function k = calc_conductivity(T)
    %Stainless steel
    k = 14.6 + 1.27e-2 * T;
end
```

- getFaceTemperature(I, j, T, T1, T2, face): which returns the correspondent temperature of a face based on the cell centroids temperatures and boundary conditions:

```
function Tf = getFaceTemperature(i,j,T,T1,T2,face)
    in = i;
    jn = j;
    switch face %Get neightbour cell index
        case 'N'
            in = i + 1;
```

```
        jn = j;
    case 'S'
        in = i - 1;
        jn = j;
    case 'E'
        in = i;
        jn = j + 1;
    case 'W'
        in = i;
        jn = j - 1;
    otherwise
        disp('Exception: Wrong
orientation reference')
    end

    if jn < 1 || in < 1 || jn >
size(T,2) || in > size(T,1)
        if jn < 1 %BC
            Tf = T2;
        elseif in < 1 % BC
            Tf = T1;
        else % Heat flux == 0 ->
Grad(T) = 0:
            Tf = T(i,j);
        end
    else
        Tf = 1/2 * (T(i,j) + T(in,jn));
    end
end
```

Moreover, two changes have been made to the main script:

- Include the non-linear loop as long as the convergence criteria and a plot of the convergence of the method. Note that the method is unrelaxed. (underlined in blue)
- Include the effect of conductivity as it have been neglected in the previous

version because of being constant. (underlined in green)

```matlab
% Problem definition
m = 50; % Number of elements in the x direction
n = 50; % Number of elements in the y direction
T1 = 773.15; % Temperature of the west wall
T2 = 273.15; % Temperature of the south wall
S = 1; % Cell face surface
V = 1; % Cell Volume
niter = 0; % Number of iterations
res0 = 1; % Initial residual, for scaling purposes

% Temperature matrix initialization and
transformation into vector

T = 0.5*(T1+T1)*ones(m,n); % Important for the
non-linear sim
Tvector = reshape(T, [m*n 1]); % Used for future
versions of the script
Tvector_new = Tvector;

% Solution matrix assemble
%M = spalloc(m*n, m*n, 4*m*n); % Allow 4
elements per row to be non-zero
M = zeros(m*n);
b = zeros(m*n,1); % independent term vector
conv = false;
while ~conv
    disp('Mattrix Assembly');
    tic
    for i = 1 : size(T, 1)
        for j = 1 : size(T, 2)
            row = sub2ind(size(T), i, j);




            % North Face

            M(row,:) = M(row,:) + 1/2
```

```matlab
                *calc_conductivity(getFaceTemperature(i,j,T,T1,T2
,'N'))*(calcGrad(i, j, M, T, S, V, 2) +
calcGrad(i+1, j, M, T, S, V, 2));
                b(row) = b(row) + 1/2 *
(calcIndependentTerm(i, j, S, V, T, T1, T2) +
calcIndependentTerm(i+1, j, S, V, T, T1, T2));

            % South Face
            if i == 1
                M(row,:) = M(row,:) -
calc_conductivity(T(i,j))*(calcGrad(i, j, M, T, S, V,
2));
                b(row) = b(row) -
(calcIndependentTerm(i, j, S, V, T, T1, T2));
            else
                M(row,:) = M(row,:) - 1/2 *
calc_conductivity(getFaceTemperature(i,j,T,T1,T2,'
S'))*(calcGrad(i, j, M, T, S, V, 2) + calcGrad(i-1,
j, M, T, S, V, 2));
                b(row) = b(row) - 1/2 *
(calcIndependentTerm(i, j, S, V, T, T1, T2) +
calcIndependentTerm(i-1, j, S, V, T, T1, T2));
            end
            % East Face
            M(row,:) = M(row,:) + 1/2 *
calc_conductivity(getFaceTemperature(i,j,T,T1,T2,'
E'))*(calcGrad(i, j, M, T, S, V, 1) + calcGrad(i,
j+1, M, T, S, V, 1));
                b(row) = b(row) + 1/2 *
(calcIndependentTerm(i, j, S, V, T, T1, T2) +
calcIndependentTerm(i, j+1, S, V, T, T1, T2));

            % West Face
            if j == 1
                M(row,:) = M(row,:) -
calc_conductivity(T(i,j))*(calcGrad(i, j, M, T, S, V,
1));
                b(row) = b(row) -
(calcIndependentTerm(i, j, S, V, T, T1, T2));
            else
                M(row,:) = M(row,:) - 1/2 *
```

```matlab
calc_conductivity(getFaceTemperature(i,j,T,T1,T2,
W'))*(calcGrad(i, j, M, T, S, V, 1) + calcGrad(i, j-
1, M, T, S, V, 1));
                b(row) = b(row) - 1/2 *
(calcIndependentTerm(i, j, S, V, T, T1, T2) +
calcIndependentTerm(i, j-1, S, V, T, T1, T2));
            end
        end
    end

    toc

    % Cuthill-Mckee
    if niter == 0 % Only calculated once for saving
resources
        p = symrcm(M);
    end
    disp('Solving with Cuthill-McKee...');
    tic
    Tvector_new(p) = M(p,p) \ b(p);

    toc

    T = reshape(Tvector_new, size(T));
    % Limiters
     T(T>T1)=T1;
     T(T<T2)=T2;


    if niter == 0
        res0 = max(Tvector-Tvector_new);
        res = 1;
    else
        res = [res max(Tvector-Tvector_new)/res0];
    end
    Tvector = Tvector_new;
    niter = niter + 1;

semilogy(linspace(1,length(res),length(res)),res)
    drawnow
```

```matlab
    if res(end) < 0.01 % Stopping criteria
        conv = true;
    end

end
%Visualization

figure
imagesc(flipud(T))
colorbar
```

The function calc_independent_term has also been modified to account the conductivity:

```matlab
function independentTerm =
calcIndependentTerm(i, j, S, V, T, T1, T2)
    %
    % Get the linear index from an element of a 2D
matrix
    %

    independentTerm = 0;

        if j == 0 || j == 1 % Diritchlet condition left
wall
        % Gradient must be continous throught the
wall so
        % the gradient of the ghost cell is the same
as the first cell
            if i < 1
                independentTerm = independentTerm +
calc_conductivity(T(1,1))*T2*S/V;
            elseif i > size(T,1)
                independentTerm = independentTerm +
calc_conductivity(T(size(T,1),1))*T2*S/V;
            else
                independentTerm = independentTerm +
calc_conductivity(T(i,1))*T2*S/V;
            end
        end
```

```matlab
    if i == 0 || i == 1 % Diritchlet condition
lower wall
        % Gradient must be continous throught the
wall so
        % the gradient of the ghost cell is the same
as the first cell


        if j < 1
            independentTerm = independentTerm +
calc_conductivity(T(1,1))*T1*S/V;
        elseif j > size(T,2)
            independentTerm = independentTerm +
calc_conductivity(T(1,size(T,2)))*T1*S/V;
        else
            independentTerm = independentTerm +
calc_conductivity(T(1,j))*T1*S/V;
        end

    end


end
```

The problem with NS equations: the SIMPLE algorithm

The previous sections have allowed the reader to get some understanding on how the equations are solved in a CFD solver.

However, when the fluid flow is meant to be obtained and not only the heat transfer equation, a new problem arise.

Till now, with the energy equation, every equation was associated with only one fluid flow variable. In the case of the Navier Stokes equations the variables to solver are the three components of the velocity and the pressure (when not adding the energy equation).

The momentum equation provides an equation for each of the velocity components. However, the pressure is not uniquely represented in a single equation.

The solution of the complete system of equations would lead to highly sparse matrixes, slowing the iterative procedure.

A solution was found by Patankar in [3] by decoupling the solution of speed and pressure using the so-called SIMPLE algorithm that can be described as in **Error! Reference source not found.** and explained in the following:

- First step: Employ an estimation of the pressure field.

- Second step: Solve the momentum equation for obtaining initial speed guesses.

- Third step: Correct the initial guess for the pressure employing the continuity equation and the calculated speeds.

- Fourth step: Recalculate the speeds with the corrected pressure.

- Fifth step: The remaining equations as turbulence, species or energy are calculated.

- Sixth step: The convergence is checked, if not satisfied the current pressure field is used as initial guess for the next iteration.

Turbulence modelling

The previously introduced equations are valid for solving any fluid flow problem. However, the non lineal term $\rho u \cdot \nabla u$ provokes a particular behavior in certain conditions known as turbulence. This phenomena is key part of almost every industrial fluid related application, the external flow around a car, aircraft design, etc...

From the antiquity scientifists have studied turbulent flows. For example in Figure 0.32 a drawing from Leonardo da Vinci is shown where he described the different size eddies that appeared in a waterfall. The complexity and different scales are two key characteristics of turbulence as will be further developed in the following.

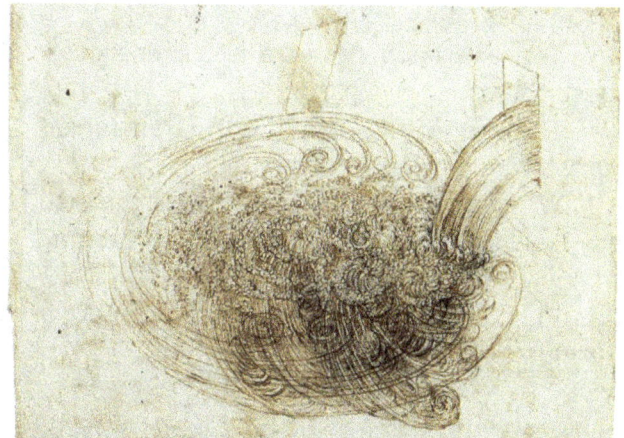

Figure 0.32 Drawing from Leonardo da Vinci showing the different turbulent scales in a waterfall.

Other characteristics of turbulent flows are the following:

- Irregularity, randomness and chaos.
- Different spatial and temporal scales.
- The higher the Reynolds number the smaller the size o f the smallest eddies.
- Fast fluctuation of all fluid variables in time.
- 3D flows although the averaged solution may be 2D.
- Coherent structures can be appreciated and are repeated in different types of flows.

Energy cascade

Already in the 1940s, Andréi Kolmogorov observed some important features of turbulent flows. Among the different scales previously mentioned, those eddies of smaller size had the greatest vorticity whilst the largest eddies had the greatest energy. From this concept came the so-called energy cascade of Kolmogorov. His important contribution can be briefly summarised in the following statements:

- The averaged flow transfers energy to the biggest eddies.
- The energy is then transferred to progressively smaller eddies.
- When the Kolmogorov scale is reached, the energy is dissipated as heat by viscous heating.

This energy cascade is represented in the following Figure:

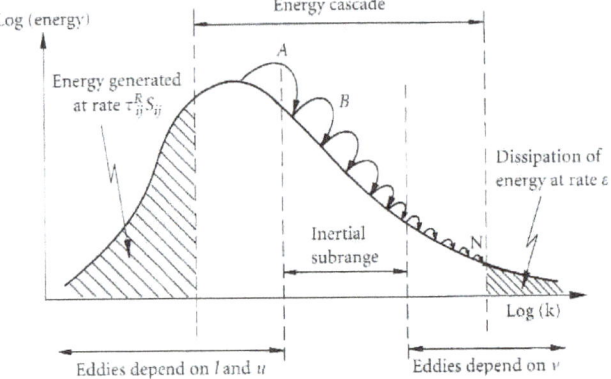

Figure 0.33 Illustration of the energy cascade,
$k = \frac{2\pi}{\lambda}$

Turbulent vs Laminar flow
The previously mentioned characteristics of turbulent flows translate in important differences when compared to laminar flows, especially in phenomena related with diffusivity and boundary layer.

Diffusivity:

- The diffusivity is increased, increasing the mixing rates.

- The energy exchange rates are enhanced as long as the momentum exchange in the walls.

Boundary layers:

- Delayed boundary layer separation.
  - Concept behind the use of vortex generator in aircrafts or the design of tennis and golf balls.

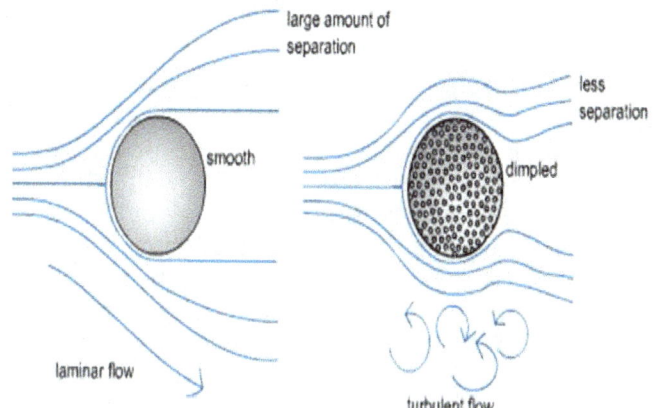

Figure 0.34 Differences between laminar and turbulent boundary layer in the flow around a sphere.

Inviability of the direct simulation

The main characteristics of the turbulent flows have been explained but one question arise how do this affect the equations? As explained earlier the problem comes from the convective term. The transition when this phenomena is expected to occur is dictated by the Reynolds number that relates the relative weight of the unstabilising dynamic forces and the stabilizing viscous forces:

$$Re = \frac{\rho v L}{\mu}$$

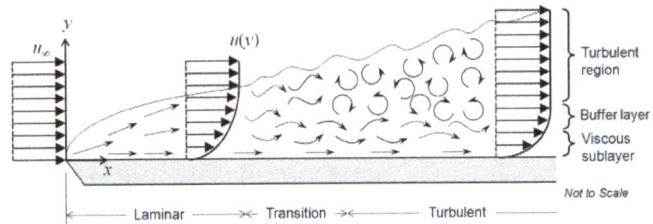

This transition can be easily observed in the Figure above where the flow over a flat plate is represented.

From what value of the Reynolds number can a flow be considered as turbulent? It depends on the geometry but a Reynolds number greater than 10^5 guarantees a turbulent flow in the vast majority of the situations. Values of Reynolds slightly below this border determine a transition region that will condition the way the near wall modelling is approached as will be later explained.

But if the previously presented equations recover all types of fluid problems why should we change the way we solve the fluid flow? For illustrating the subjacent problem the following example is employed:

Figure 0.35 2D Section, width 3L.

Considering the previous 2D section of the domain, firstly the Reynolds number is calculated:

$$Re_L = \frac{vL}{v} = \frac{94.44 \cdot 5.3}{1.544 \cdot 10^{-5}} = 3.24 \cdot 10^7$$

The Reynolds number of the fluctuating part of the speed can be approximated as:

$$Re_{u'} \sim 1\% Re_L = 3.24 \cdot 10^5$$

Kolmogorov related the smallest scales with the integral scale that is approximated with our characteristic length. More information is available [9].

$$\Delta x \leq \eta \sim \frac{l \sim L}{Re_{u'}^{\frac{3}{4}}} = 3.9 \cdot 10^{-4} m$$

Hence, the number of cells:

$$N_x = \frac{5.5L}{\Delta x} = \frac{29.15}{3.9 \cdot 10^{-4}} \sim 74800$$

$$N_y = \frac{3H}{\Delta x} = \frac{3}{3.9 \cdot 10^{-4}} \sim 7700$$

$$N_z = \frac{3L}{\Delta x} = \frac{15.9}{3.9 \cdot 10^{-4}} \sim 40800$$

$$N_{total} = N_x \cdot N_Y \cdot N_z = 2.35 \cdot 10^{13} \, cells$$

That mesh would require a RAM memory of the order of thousands of terabytes.

Next, the time step can be approximated by estimating the fluctuating speeds as 1% of the mean ones.

$$\Delta t \leq \frac{\Delta x}{u'} \sim \frac{\Delta x}{1\% u} = \frac{3.9 \cdot 10^{-4}}{0.01 \cdot 94.44} = 4.13 \cdot 10^{-4}$$

Which for solving 20 seconds of fluid flow yields:

$$N_t = \frac{20 \, s}{\Delta t} \sim 48500 \, time \, steps$$

These numbers for time steps and mesh size make completely impossible to solve these problems even in the most powerful supercomputers where some simple geometries are starting to be studied.

Turbulence models

As it has been introduce, in most of industrial and motorsport fluid dynamic problems a turbulence model is required. There are several types which are illustrated in the following Figure:

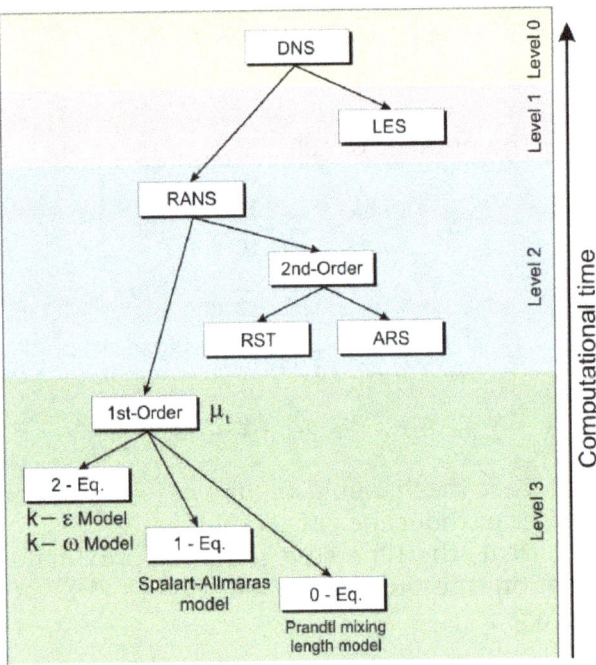

The DNS (Direct Numerical Simulation) represents the solution of all the scales without the inclusion of any modelling.

Moving downwards in the graph, the computational time is reduced and the ammount of scales modelled are increased. In most industrial applications, the 1st-Order RANS models are the state of the art and will therefore be the main focus of this section.

RANS models

As mentioned in the initial discussion about turbulence, one of its main characteristics are the fluctuation of the fluid variables as represented in. Figure 0.36.

The RANS models are named due to the averaging that is made to the fluid variables, decomposing them into an average component and a fluctuating component.

It is important to note that this averaging process may difer when the problem implicates varying density, employing the so-called Favre averaging for some variables.

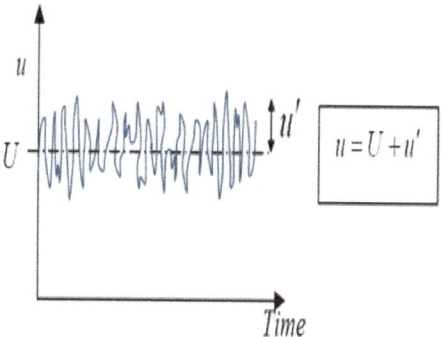

Figure 0.36 Time history of fluid velocity in a typical turbulent problem.

By employing these models some of the time scales are lost as represented in the following Figure as only the averaged component will be solved.

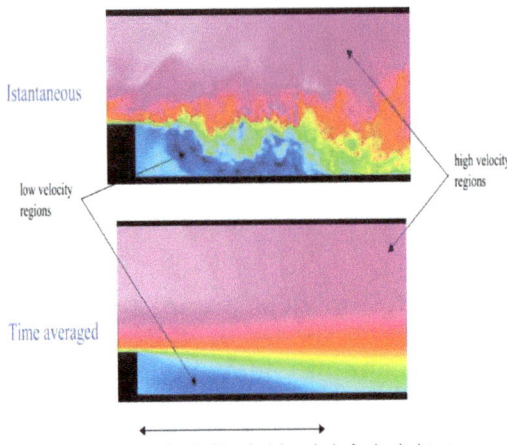

Figure 0.37 Comparison between real phenomena and RANS solution.

This averaging has implications in our governing equations as will be explained in the following. Coming back to the originally proposed governing equations for the continuity equation and the momentum equation:

$$\frac{\partial}{\partial t} \int \int \int_V \rho \, dV + \int \int_S \rho u \cdot n \, dS = 0$$

$$\frac{\partial}{\partial t} \int \int \int_V \rho u \, dV + \int \int_S \rho u (u \cdot n) \, dS$$
$$= \int \int_S PI n \, dS + \int \int_S \overline{\overline{\tau^i}} \, n \, dS + \int \int \int_V f_v \, dV$$

Now, every fluid variable is decomposed in the averaged and fluctuating term (for example $u = \langle u \rangle + u'$). Then, every term is averaged yielding to:

$$\frac{\partial}{\partial t} \int \int \int_V \rho \, dV + \int \int_S \rho \langle u \rangle \cdot n \, dS = 0$$

$$\frac{\partial}{\partial t} \int \int \int_V \rho \langle u \rangle \, dV + \int \int_S \rho \langle u \rangle (\langle u \rangle \cdot n) \, dS$$

$$= \int \int_S \langle P \rangle I \, n \, dS + \int \int_S (\langle \overline{\overline{\tau^*}} \rangle + \overline{\overline{\tau^R}}) \, n \, dS$$

$$+ \int \int \int_V f_v \, dV$$

At first glance, the equations looks fundamentally the same and most of the fluctuating components vanished. However, during the decomposition of the stress tensor a term appears $(\overline{\overline{\tau^R}})$ that is dependent on the fluctuating components. This term is the so-called Reynolds stress tensor and can be xpressed as:

$$\overline{\overline{\tau^R_{ij}}} = \rho \langle u_i u_j \rangle$$

As the idea behind the averaging is avoiding the necessity of solving these fluctuating terms, a model must be employed for accounting for this tensor. The way this tensor is modeled difer in the different RANS models. In the models some of the approaches will be briefly explained. For a deeper knowledge, the reader is advised to consult the wide bibliography provided.

First Order Models

All of them are based in the Boussinesq Hypothesis that states that the momentum exchange between eddies occurs in a similar fashion as the one between molecules due to viscosity.

Following this assumption, which not even Boussinesq was sure about it but has been shown to work effectively, the Reynolds tensor is approximated by using a turbulent eddy viscosity.

The differences between the turbulent models lie in how this artificial viscosity is calculated.

Prandtl mixing length model

The model is based in a concept similar to the mean free path of the kinetic gas theory, stating the fluctuating components as:

$$u'_i = -l_{mixing} \frac{\partial \langle u_i \rangle}{\partial \eta}$$

From where the components of the Reynolds tensor can be calculated as:

$$\langle u_i u_j \rangle = l_{mixing} \left| \frac{\partial \langle u_j \rangle}{\partial \eta} \right| \frac{\partial \langle u_i \rangle}{\partial \eta}$$

Hence:

$$\nu_t = l_{mixing} \left| \frac{\partial \langle u_j \rangle}{\partial \eta} \right|$$

The mixing length is a constant parameter that highly depends on the problem.

However, on this model are based the wall functions that play an important role and will be explained later so it is important to have a basic understanding.

Figure 0.38 Illustration of the mixing length concept.

In order to understand the mixing length concept the example of the previous Figure will be taken as starting point. In it the mean temperature is represented as a colourmap. If a portion of the fluid, as represented in the left part of the illustration, moves to a region where the mean temperature is different, the temperature of the portion of fluid does not change immediately if the scale is small enough. We can then express the temperature in the region 1 (botton) and 2 (top) as:

$$T_1 = \langle T_1 \rangle$$
$$T_2 = \langle T_2 \rangle + T_2{'}$$

As we have assumed that the fluid temperature does not change, we can say:

$$\langle T_1 \rangle = \langle T_2 \rangle + T_2' \Rightarrow T_2' = -(\langle T_2 \rangle - \langle T_1 \rangle) = -l_{mixing} \frac{\partial \langle T \rangle}{\partial z}$$

Being l_{mixing} the length in which the fluid does not experiment a temperature change. Prantl employed this concept applied to the turbulent flows obtaining the previously presented equation:

$$u'_i = -l_{mixing} \frac{\partial \langle u_i \rangle}{\partial \eta}$$

Spalart-Allmaras

Introduces eight empirical coefficients and three closure functions. Defines the eddy viscosity as:

$$\nu_t = \tilde{\nu} f_{v1}$$

For obtaining the new variable $\tilde{\nu}$, a new differential equation is added which is here presented in the differential form employing Einstein's notation:

$$\frac{\partial \tilde{\nu}}{\partial t} + \langle u_j \rangle \frac{\partial \tilde{\nu}}{\partial x_j} = c_{b_1} \tilde{S} \tilde{\nu} - c_{w_1} f_w \left(\frac{\tilde{\nu}}{d}\right)^2$$
$$+ \frac{1}{\sigma} \frac{\partial}{\partial x_k} \left[(\nu + \tilde{\nu}) \frac{\partial \tilde{\nu}}{\partial x_k} \right] + \frac{c_{b_2}}{\sigma} \frac{\partial \tilde{\nu}}{\partial x_k} \frac{\partial \tilde{\nu}}{\partial x_k}$$

This turbulence model has been widely employed in the literature, especially in aerospace problems where the flow is expected to be attached. One of the main disadvantages is when it comes to predicting separation.

$k - \epsilon$ models

They introduce two variables, the turbulence kinetic energy (k) and its dissipation (ϵ). The eddy viscosity is defined in terms of these new variables as:

$$v_t = \frac{C_\mu k^2}{\epsilon}$$

For obtaining both k and ϵ, the following transport equations are added:

$$\frac{\partial}{\partial t}\rho k + \frac{\partial}{\partial x_i}\rho k u_i$$

$$= \frac{\partial}{\partial x_j}\left[\left(\mu + \frac{\mu_t}{\sigma_k}\right)\frac{\partial k}{\partial x_j}\right] + P_k - \rho\epsilon - \rho D$$

$$+ S_k$$

$$\frac{\partial}{\partial t}(\rho\epsilon) + \frac{\partial}{\partial x_i}(\rho\epsilon u_i)$$

$$= \frac{\partial}{\partial x_j}\left[\left(\mu + \frac{\mu_t}{\sigma_\epsilon}\right)\frac{\partial\epsilon}{\partial x_j}\right] + C_{1\epsilon}f_1\frac{\epsilon}{k}P_k$$

$$- C_{2\epsilon}f_2\rho\frac{\epsilon^2}{k} + \rho E + S_\epsilon$$

These models are known to be very robust models and not very sensitive to the inlet turbulent conditions that are usually hard to determine.

k-ω models

These models introduce two fluid variables: the turbulence kinetic energy (k) and its specific dissipation rate (ω). The eddy viscosity is defined as:

$$v_t = \frac{k}{\omega}$$

$$\frac{\partial k}{\partial t} + u_j\frac{\partial k}{\partial x_j} = \tau_{ij}^R\frac{\partial u_i}{\partial x_k} + \frac{\partial}{\partial}x_j\left[(v + \sigma^* v_t)\frac{\partial k}{\partial x_j}\right] - \beta^* k\omega$$

$$\frac{\partial\omega}{\partial t} + u_j\frac{\partial\omega}{\partial x_j} = \frac{\partial}{\partial x_j}\left[(v + \sigma v_t)\frac{\partial\omega}{\partial x_j}\right] + \alpha\frac{\omega}{k}\frac{\tau_{ij}^R}{\rho}\frac{\partial u_i}{\partial x_j} - \beta\omega^2$$

Where $\alpha, \sigma, \sigma^*, \beta$ and β^* are empiric coefficients of the model.

The advantages of these models lie in a better behavior under strong curvatures and adverse pressure gradients but they are highly dependent on inlet conditions.

SST $k - \omega$ model

In order to combine the strengths of the $k - \epsilon$ models and the $k - \omega$ model, the SST $k - \omega$ was developed by Menter [10]. This model employs a shear stress transport formulation (SST) that allows to use a $k - \omega$ approach close to the wall whilst using a robust $k - \epsilon$ in the outfield.

Wall treatment

The areas close to the walls are particulary critical and require special attention as the gradients encountered are big. That is the reason why the flow near the wall, where the boundary layer is developed, has been widely studied in the literature and some common behaviours have been observed when related to turbulent flows. As it is represented in the following Figure:

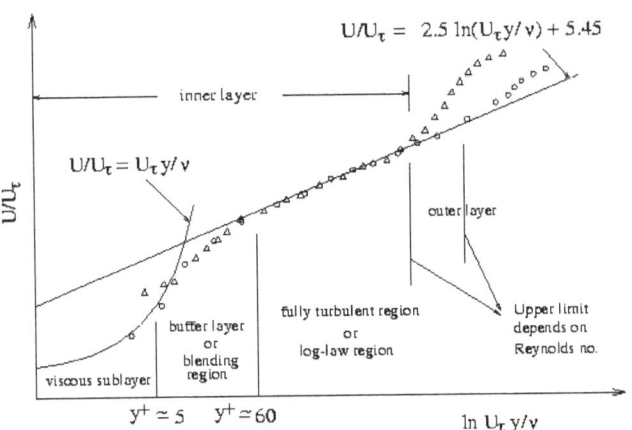

Figure 0.39 Experimental results obtained for turbulent boundary layers.

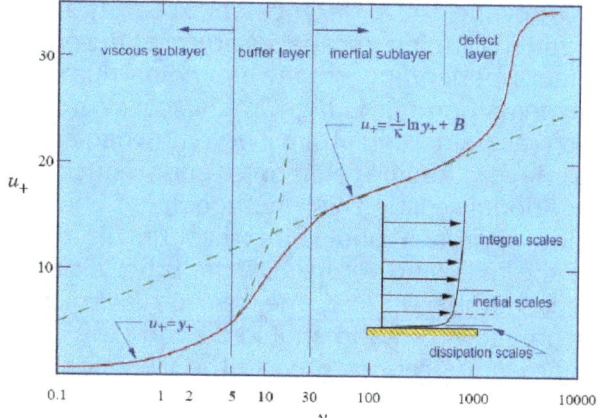

Three main areas are distinguished in terms of a dimensionless height y^+ where different profiles are consistenly obserbed for the dimensionless speed u^+. These two dimensionless numbers are key in the boundary layer theory and are expressed as:

$$u^+ = \frac{u}{u_\tau} \qquad y^+ = \frac{y \cdot u_\tau}{\nu} \quad with \ u_\tau = \sqrt{\frac{\tau_w}{\rho}}$$

Where τ_w is the stress tensor in the wall.

The nearest layer to the wall is called *viscous sublayer*, where the molecular viscosity dominates the flow behavior and the effects of the turbulent viscosity are neglible. In this region the velocity profile has been found to be $u^+ = y^+$.

The *buffer layer* is a transitional layer towards the *log-layer* which is characterized by the eddy viscosity and the velocity profile is logarithmic as anticipated by the layer name.

Knowing the differences between the layers helps understanding the necessity of different wall treatments depending on the Reynolds number. If the Reynolds number is high, the viscous forces are not so relevant and then we can neglect the viscous layer, saving elements and reducing the computational cost. This methodology is a high Reynolds approach and is represented in the left side of the following Figure. On the other side if the Reynolds number is low, the viscous layer can no longer be neglected and should be correctly solved, this approach is represented in the righ side of the Figure.

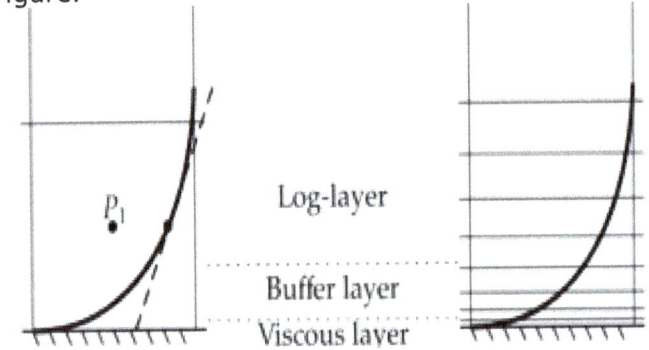

Figure 0.40 Wall treatment approaches: (left) High Reynolds (right) Low Reynolds.

The high Reynolds approach highly simplifies the problem as only the log-layer is modelled and algebraic expressions are used. These expressions are different for each turbulence model and are called wall functions. For example, for a $k - \epsilon$ model:

$$\langle u \rangle = \frac{u_\tau}{\kappa} \ln y + C, k = \frac{u_\tau^2}{\sqrt{C_\mu}}, \epsilon = \frac{u_\tau^3}{\kappa y}$$

It is important to remark that the wall funcitons are based in experiments with attached flow and although corrections exist for separated flow, they are not particulary accurate in these conditions.

Best practice guidelines

Geometry

In every CFD study, the employment of a CFD-prepared geometry takes a great amount of the total human time spent in the Project, even above 75 %.

This is an underestimated when this field is firstly approached and can be a great source of frustration.

The most common CFD format is the IGES format that is a standard between the different CAD software and allow saving surfaces (note not solids are saved). Another good alternative is the STEP format that allow the user to export solids and offers better behaviour in some software and also save the hierarchical structure of the model.

The first thing to notice is that we need to mesh only our computational domain, which is the fluid based domain. The only exception would be in those cases we are going to perform a fluid-structure iteration analysis or want to model the heat transfer inside the solids parts of our model (conjugate heat transfer). In general terms for aerodynamic problem Is crucial that a single solid is employed for described all the domain where the air flows.

The choice of the computational domain can have a great effect of results as in experiments the wind tunnel aspect ratio has. The smaller of our domain the less elements we will require but at the same time we can underestimate the drag and lift force produced.

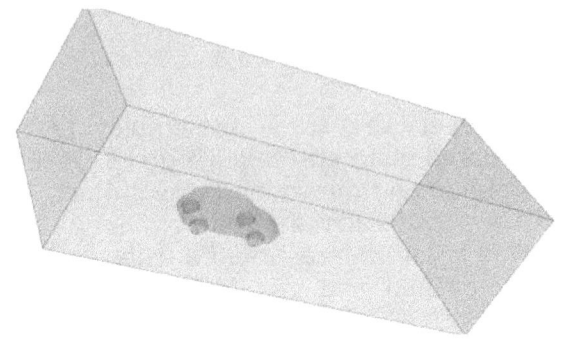

Some reference dimensions are given above, but it is very important to make a mesh analysis of the dependence of our results on the size of our computational domain to have comparable results.
Given a car length "L":

Total length of the domain: 20 L
Length upstream the obstacle: 5 L
Length downstream the obstacle: 15 L
Total height: 15 L

For a first rough analysis, the tires can be supposed connected to the chassis by just a rigid axle. The detail level, of course, will depend on the level of accuracy required. The tire-chassis connection will be more important in open-wheeler simulations that is in touring cars.

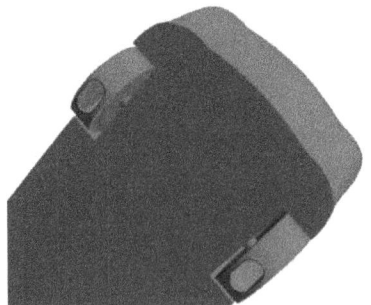

The contact surface between the tire and our virtual track can be just a line as it would be a perfect cylinder. In case of not having proper tire deformation data, a rough print must be drawn, this is crucial for achieving a proper mesh later in the pre-processing process as ensures lower contact angles in the connection between tire and track.

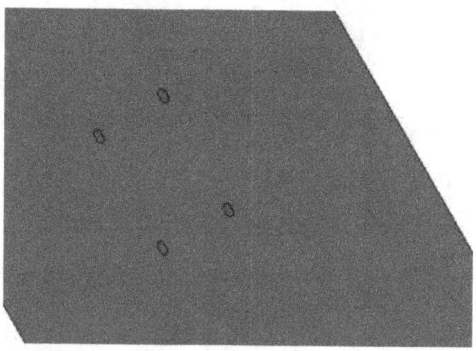

A traditional approach for a first simulation is just cutting the tire above the track and extruding the print. In this way the contact angle is 90 degrees which will make life easier to our mesher software. Again, in motorsport the detail requirements are so high that this is standard is not enough and even the iteration between the flow field and the tire deformation is closely examined.

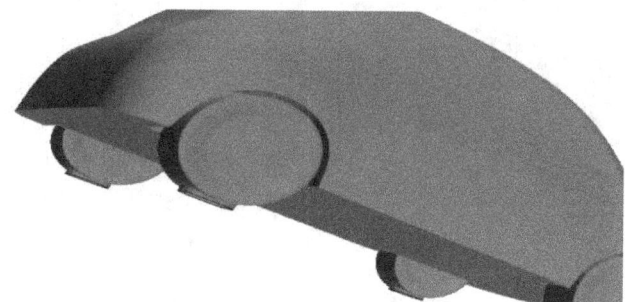

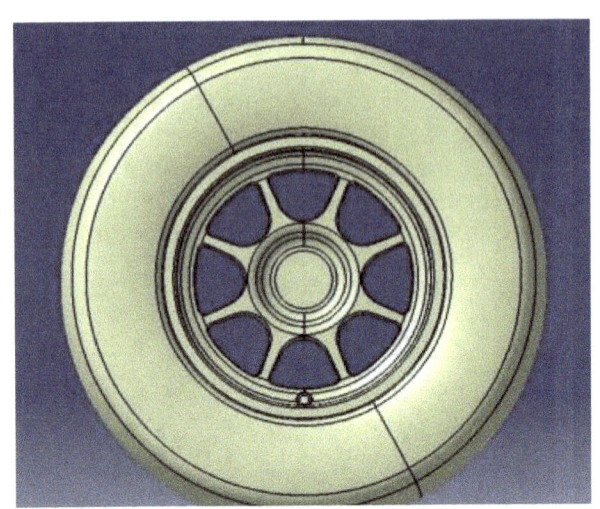

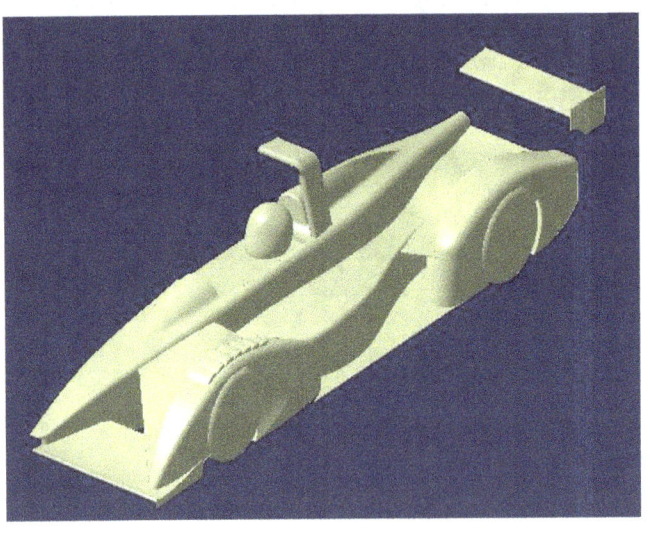

Meshing

Probably meshing is the most complicated process in a CFD Analysis and it dramatically affects the quality of the obtained results. A bad mesh could provoke the simulation to diverge or even worse, yield misleading results.

Observing the example of a 2D cylinder as the one presented in the Figure and assuming a quad mesh around it, the calculation points are the centroids of the quads in the fluid side. This limit the resolution of our solution, affecting also the level of accuracy with which the geometry is recovered.

In the example of the 2D car can be observed the usual procedure that a Cut-Cell meshing algorithm follows. A simple quad mesh is initially employed. Then, the mesh is refined around the geometry of the obstacle. Finally, a prism layer is added in the vicinity of the surface to recover the boundary layer.

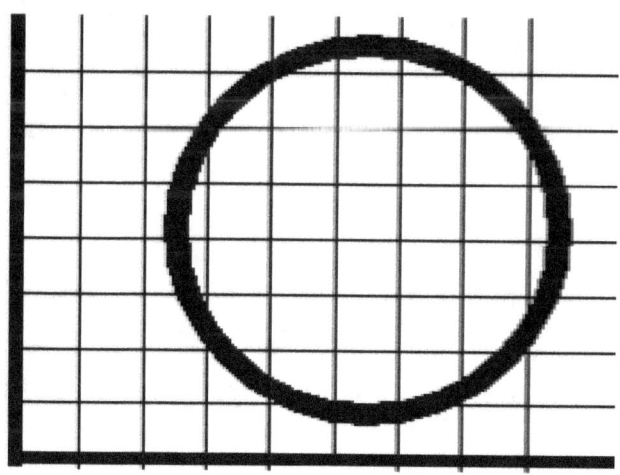

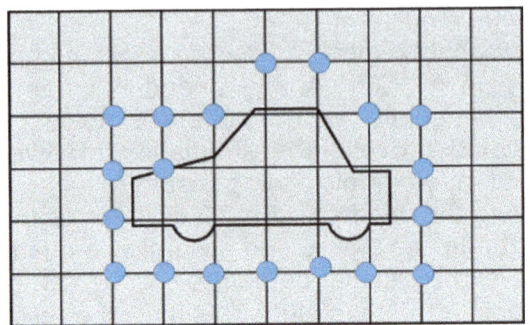

Solución: hacer la malla más y más fina; mala solución pues jamás alcanzaremos el objetivo:

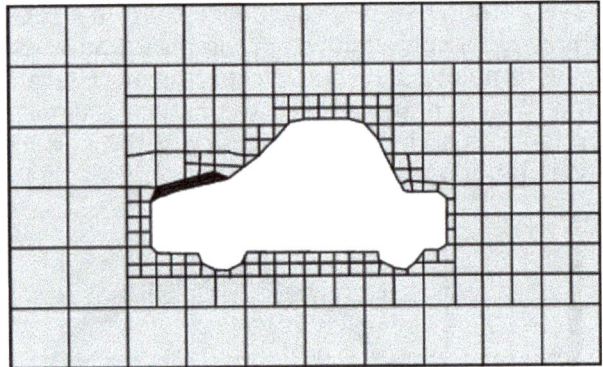

Another meshing methods are based on a structural approach and use mathematical geometrical transformations as the Riemann or the Youkowsky transformation. Examples of these are shown in the following figures.

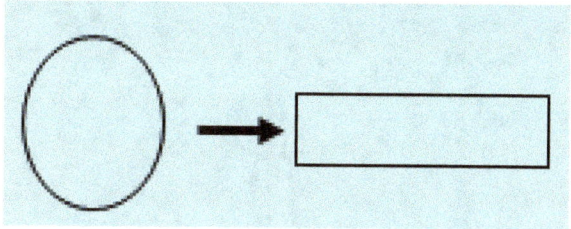

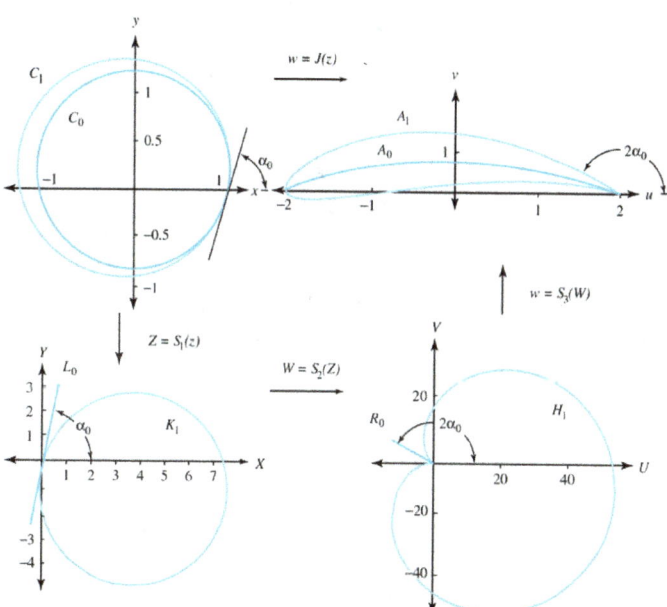

However, these methods usually require greater time and effort and are not the usual practice in such a fast-paced industry as motorsport. The alternative is the unstructured meshing where the quality of the elements must be carefully analysed in order to guarantee the validity of the solution.

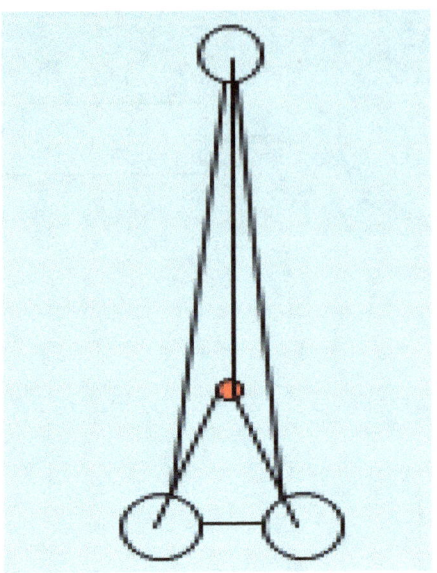

As it has been previously introduced, the discretization of the equations require performing vector operations where the normal of the faces and the vectors joining different cells are involved. Therefore, if the cell shapes are degenerated as the one showed in the upper Figure, the accuracy is compromised.

A preferred geometry, when the complexity of the problem allows it, is the hexahedra. Hexa meshes keep an angle between the face normal and the vector joining cell centroids almost parallel.

In order to evaluate the quality of the mesh elements, different mesh meters were created. All of them usually compare the element characteristics to an ideal cell.

One of the most employed is skewness, representing a value of 1 a fully degenerated cell and a value of 0 an ideal mesh. A rule of thumb would be keeping the surface maximum skewness bellow 0.8 and the volumetric skewness bellow 0.9. Modern solvers are able to cope with very deformed elements when their amount is low. For a more detailed explanation is useful to refer to the manual of the employed solver as it will contain reference values recommended. However, the reader must remember that a converged solution is not necessarily an accurate one and the mesh quality plays an important role on it.

$$\text{Skewness} = \frac{\text{optimal cell size} - \text{cell size}}{\text{optimal cell size}}$$

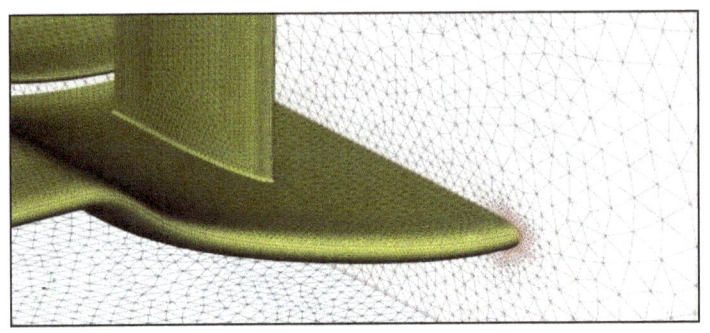

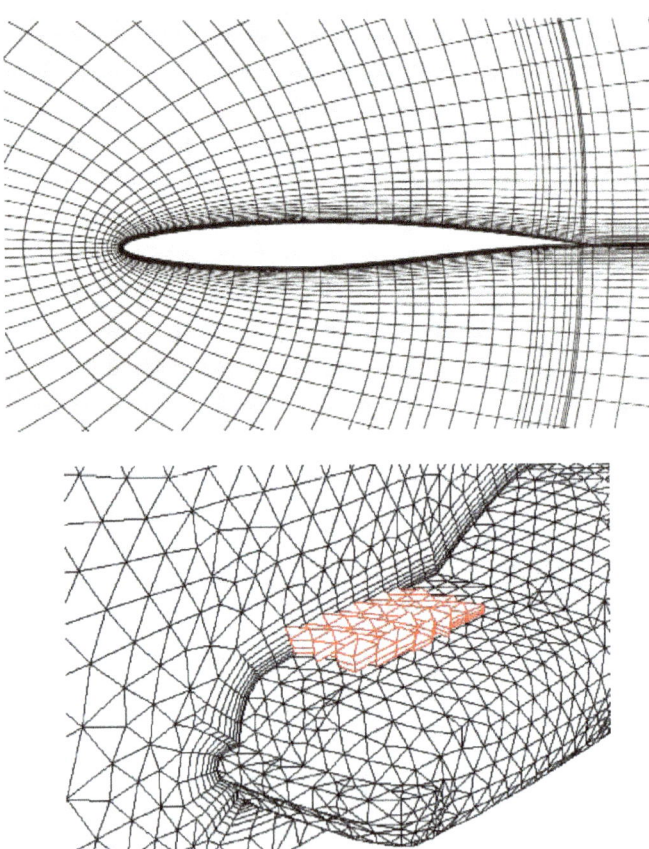

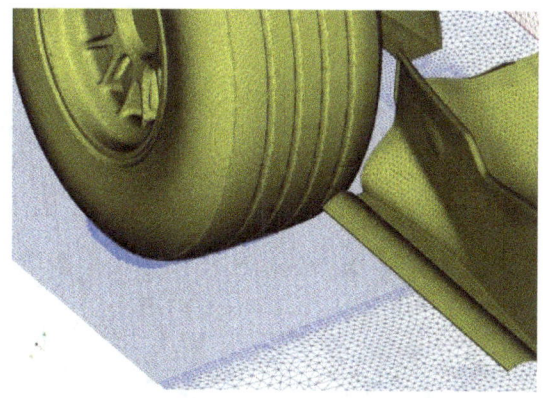

The most critical step for obtaining a high quality mesh is providing the meshing algorithm with a geometry as clean as possible. A clean geometry refers to a closed surfaces collection where the number of surfaces is limited to the ones necessary to describe with a satisfactory level of accuracy the analyzed geometry. Examples of meshes of incorrect geometries can be found in the following figures:

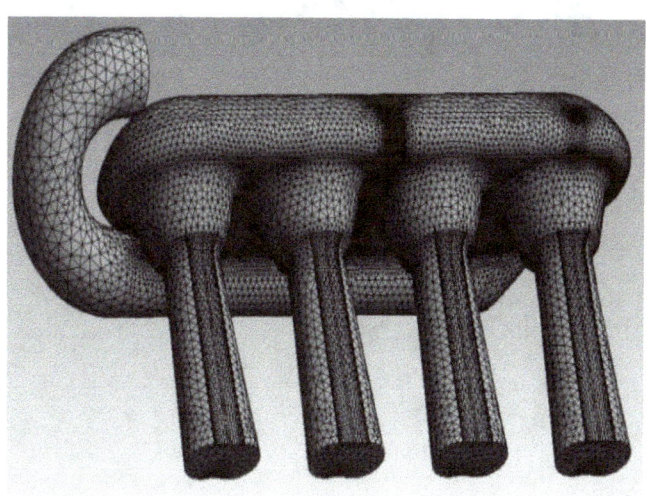

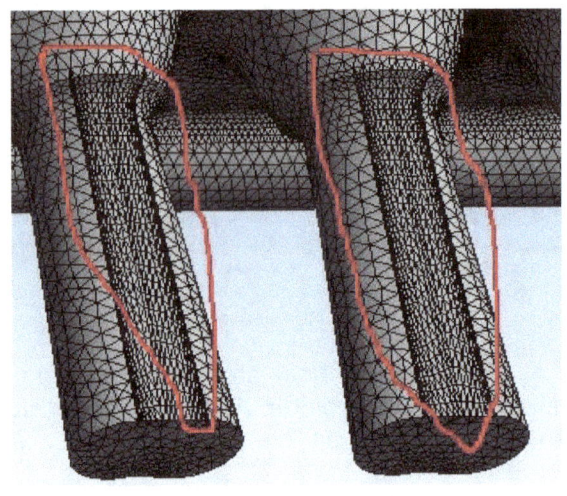

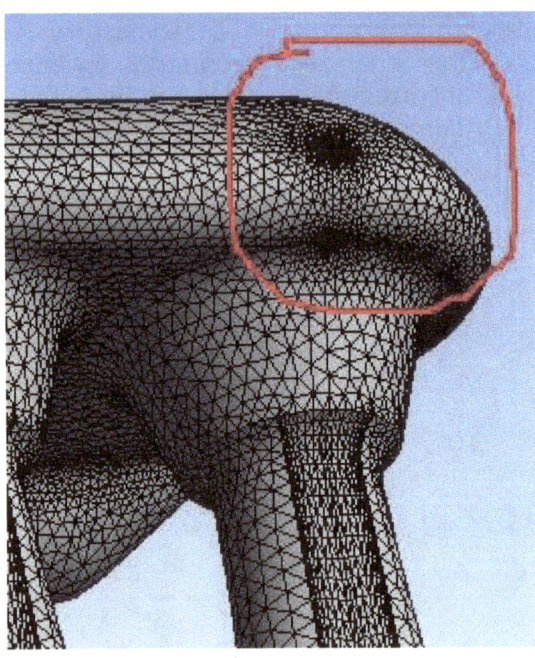

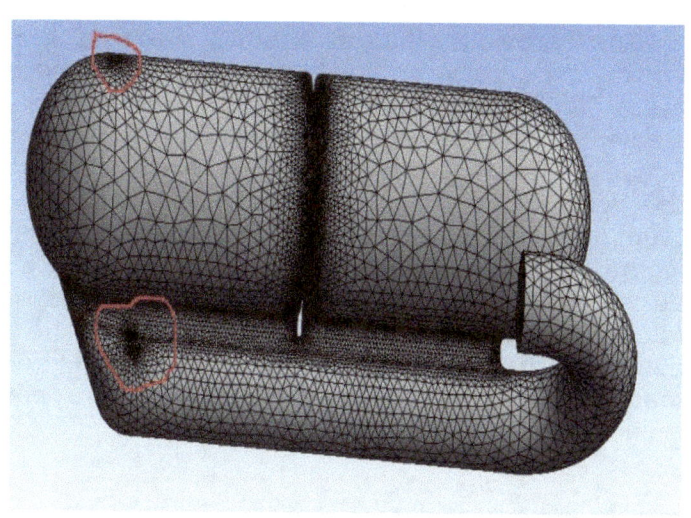

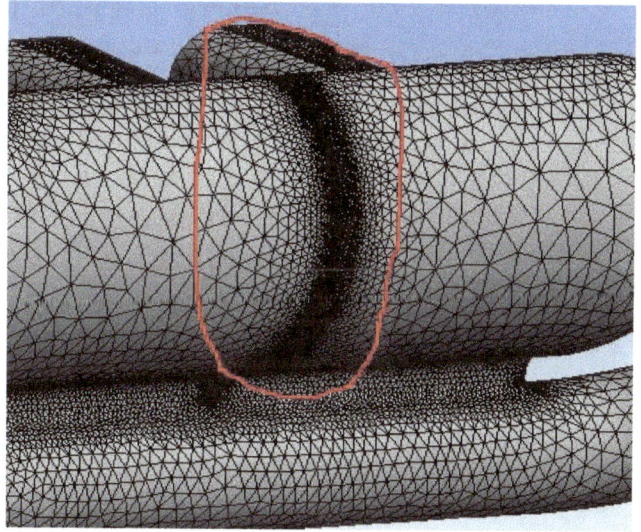

The previous figures show meshes with unexpected areas of mesh concentration. These areas limit the capability of spending more elements in those features that have a greater impact in the fluid phenomena and hence damage the accuracy of the solution. The reason behind them may be an unintended line as in the last Figure or very small degenerated surfaces created during the geometry reconstruction. Modern meshers have defeaturing capabilities that help avoiding these problems by automatically or manually selecting those elements that are not required to describe the geometry.

Solver setting

This is maybe the most standard part of the simulation and usually does not require a great effort as aerodynamic simulations are similar in nature. However is very important to have ground fundamentals about turbulence modelling in order to have reliable results.

Once a certain CFD software is learnt, the steps are very similar for every of them so the above is not directed to a concrete CFD suite. In the following some useful guidelines are given.

It is important to define early on what we want to get form our simulation.

This is very important specially in transient simulations where the saving frequency hast be way lower than the time steps and the fluid flows frequencies and it would be necessary to repeat the simulation with the associated cost in hours and money. It is also important to divide the geometry in useful groups in order to make the post-processing easier, for example creating a separated group for the rear wing if we are interested in checking its efficiency. As it has been mentioned earlier, every step we take early in the process to favour the following process will be worthy.

Initial and boundary conditions:

The initial conditions are very important not only for transient flows but also as the initialization of steady runs. In the last cases is usual better to start with values of velocity close to the inlet in order to favour a faster convergence rate but there are cases where this can cause the turbulence variables to blow up, this can also be controlled by tweaking the under relaxation values, so again this is very dependent on the simulation we are facing.

Regarding an external aerodynamic simulation, the only thing we need we have to define are the velocity characteristics (and turbulence, see next section) in the inlet, define a pressure outlet, define the solid (car) and moving walls (track). It is also necessary to define the symmetry planes if employed and also the pressure loses in devices as radiators.

Turbulence models:

There are a great spectrum of turbulence models and they are in continuous developing, here a rough guidelines are given based of the explained in the theory section. The final choice is greatly influenced by the computational resources available and the level of accuracy required.

The best model for external flows in terms of best balance between problem description and computational cost is the k-w SST model. This model behaves reasonably well in separated zones that are very common in aerodynamic flows, however as every boussinesq-based model it assumes isotropic turbulence with all its consequences.

A traditional "light-weight" model for attached aerodynamic simulations is the Spalart-Allmaras model and can be used for very fast simulations in order to obtain a first design, for example inside an optimization algorithm.

In the following picture, a comparison of several turbulence models can be observed:

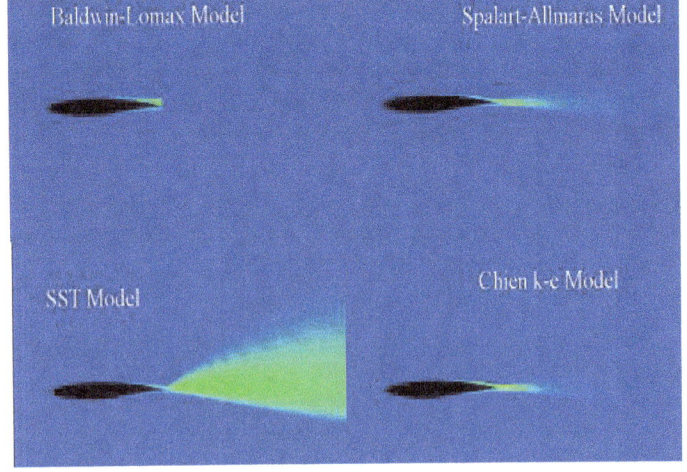

It would be ideal to contrast our model with track data if available or with relevant published data as the ones shown below for a NACA profile.

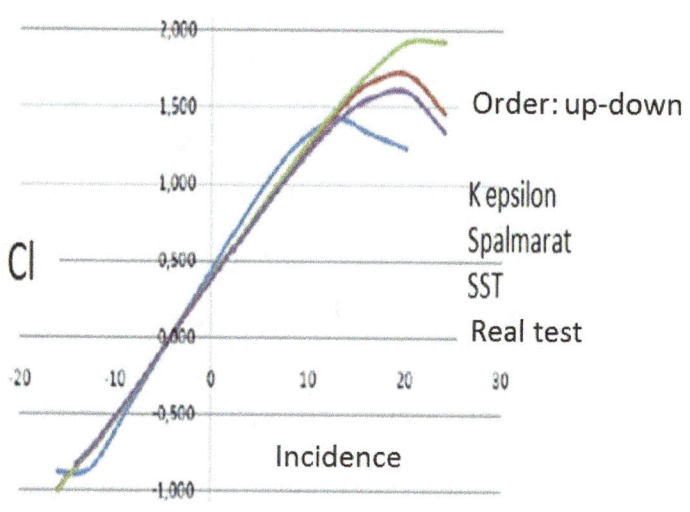

Order: up-down

K epsilon
Spalmarat
SST
Real test

Incidence

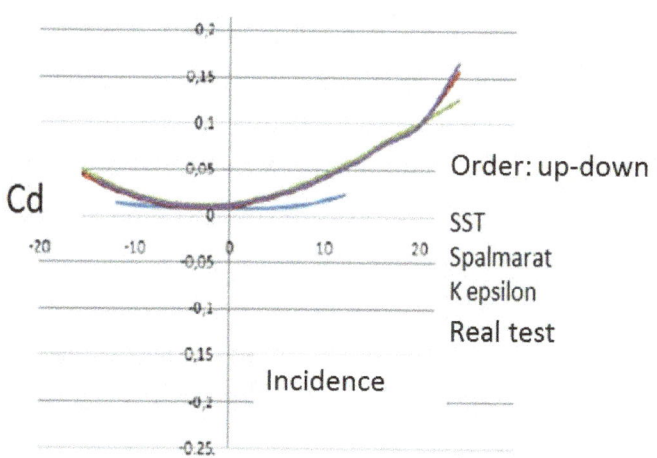

Order: up-down

SST
Spalmarat
K epsilon
Real test

Incidence

It is critical to know the limitations of the chosen turbulence model and know where its behaviour may be compromised, as for example in the possible separation points or those phenomena where a high level of swirl and shear is present. This will help avoiding taking the decisions driven by a non-physical behaviour induced by the chosen fluid models.

As a rule of thumb, the SST k-w model has been found as one of the most accurate RANS model in terms of recovering the separation and wake profile that are the most critical aspects when analysing an external aerodynamic problem.

Examples of turbulence models comparison are easily found in the literature. In [11], for example, different turbulence models were analysed regarding the solution of a Tyrell wing section.

Time discretisation:

If an explicit solver was chosen, the maximum time step will be determined by the Courant number. On the other hand, if for reducing the computational times an implicit solver was selected, is important to consider the following points:
The typical Strouhal number expected in the flow that will give the expected periods as:

$$T = \frac{L}{U \, St}$$

Sufficient time resolution must be selected in order to recover these oscillations.

If the convergence criteria is not met is less than 20 iterations is always preferable to reduce the time step, achieving more time resolution with an equivalent computational effort.

Postprocessing

In this step of the process the objectives are the following:

- Check the convergence of our run, guaranteeing the validity of the results.
- Understand the flow behavior by visualizing pathlines and different plots of the flow variables.
- Calculate meaningful values as forces or pressure drops.

A usual way of checking convergence is by setting absolute values of the residual monitors, however flow variables should be analyzed and should be stable before considering a solution as converged. Another convergence meter is the mass imbalance that can be evaluated as the difference in percentage of the mass flow entering and leaving the domain.

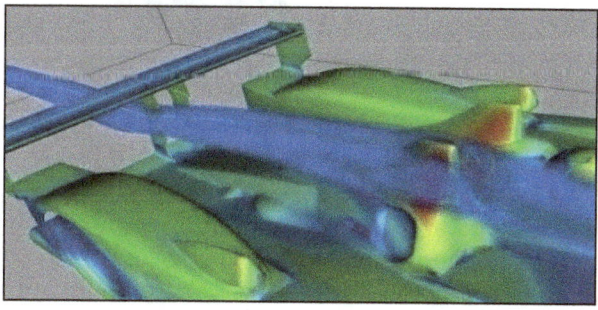

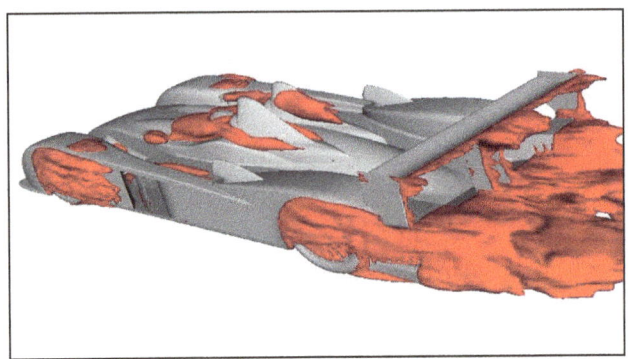

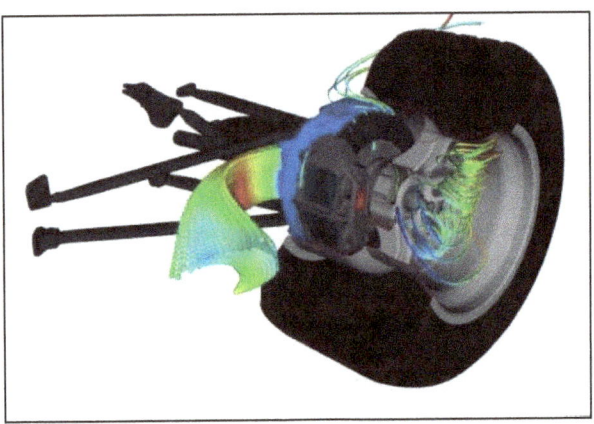

Usually the detailed analysis of the flow field is restricted to those cases were a new flow behavior is sought. For example, representations of skin friction over surface can help determining if separation occurs in a new guide vane. On the other hand, if performance in terms of downforce and drag is aimed, the values of the drag and downforce coefficients must be obtained as long as the position of the center of pressure.

Cases of interest

Airbox design and intake modelling

An important analysis in the motorsport world is the flow in the airbox. By expanding efficiently the flow, gains in power can be achieved.

One of the biggest challenges in these simulations is the modelling of the boundary conditions in the cylinder inlets.

A simplified approach would be to apply a sine wave on the mass flow admitted by each cylinder with the appropriate phase delay between them.

Another approach would be employing real data from a dyno or data obtained from a 1D Simulator as AVL Boost or GT Power. However, if the details of the flow inside the airbox are not needed, a simplified approach applying a speed-dependent mass flow can be followed and a similar approach can be adapted for the exhausts.

Another important boundaries where additional data is required are the radiators.

A common practice is modelling them as porous media, applying a certain pressure drop. Again, the required information to feed this model may come from experimental analysis or from an isolated CFD simulation of the radiator where the pressure drop is calculated.

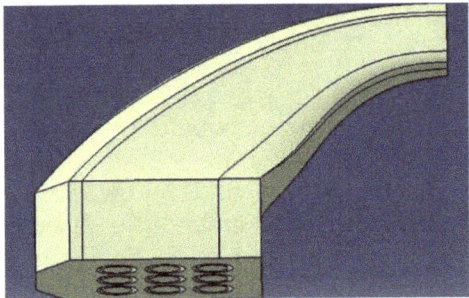

➔ <u>Resumen; conceptos e ideas a seguir:</u>

Internal Flow

Although most of the effort of this chapter has been put in the modelling of external flows, internal flows are becoming more and more important in motorsport, especially those related with thermal management.

The first step consists in extracting the outbounding surfaces of the fluid domain from a general part. This process is shown in the following Figures. Most modern meshers provide tools that assist the user in this process.

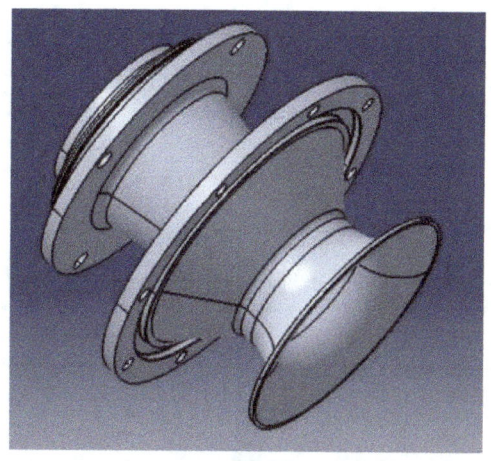

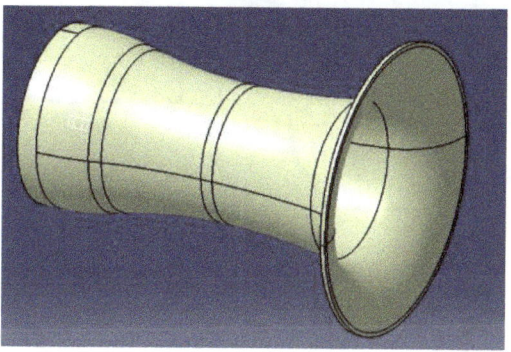

However, if the previously presented geometry were simulated, convergence problems are likely to occur. This is due to the proximity of the boundary conditions to areas of strong gradients. A usual procedure to favour convergence is to extend both inlet and outlet creating the so-called pre-inlet and post-outlet In this way a better-posed problem is generated but the final solution is also affected.

In order to address this problem a more realistic setup as the one sketched in the last Figure can be adopted. This account for the high gradient problems and at the same time represents a more accurate configuration that will recover better the flow behavior.

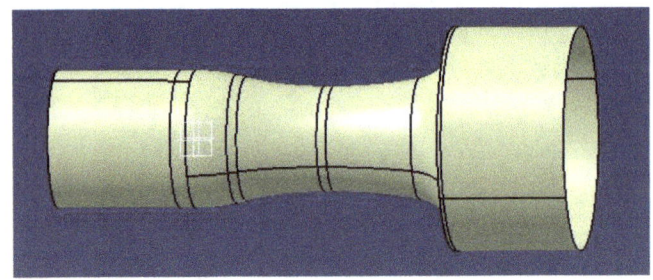

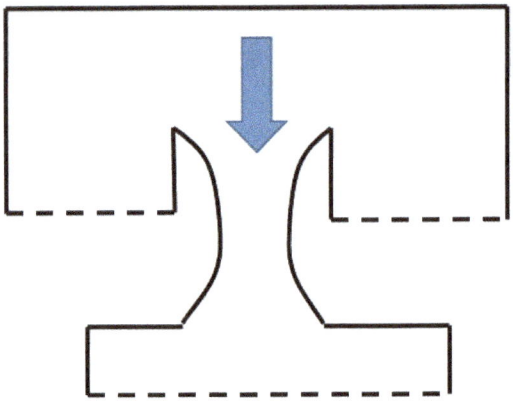

Tyre geometry

The flow around the tyres is probably one of the biggest headaches when track-simulation correlation is concerned. An accurate geometrical representation helps to close the gap. The tyre generates three vortex on each side, one in the contact patch, another one approximately in the centre line and one in the top edge.

These three vortexes dramatically influence the global aerodynamic performance and a small deviation may change strongly the car behaviour.

The shape of the tyre continuously changes with the vertical load and due to the suspension geometry, making it a very complex problem. Some tables previously introduced in the book give an idea of the amount of deflection expected. An example of a tyre under different loads can be found in the following Figure, where the green contour represents an unloaded tyre.

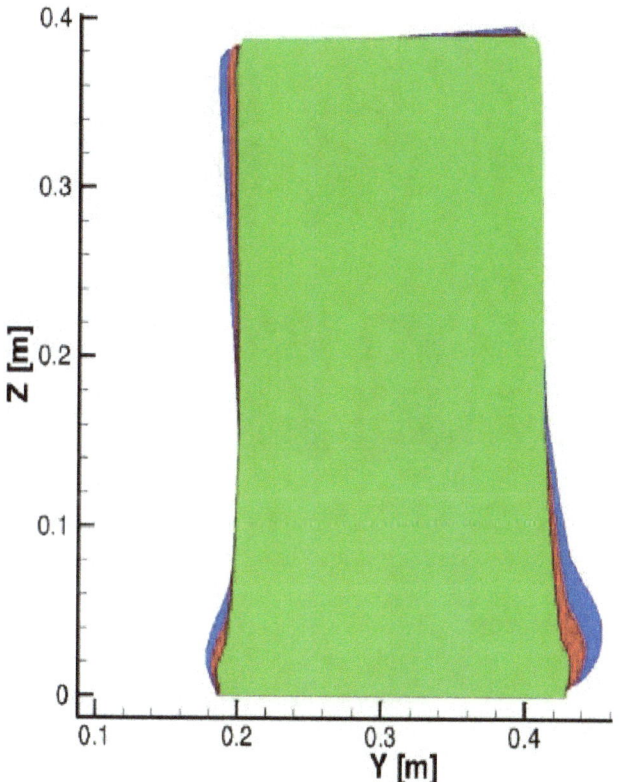

A further more detailed analysis is found in the following Figure, where the camber effects are analysed. Both Figures are extracted from [12].

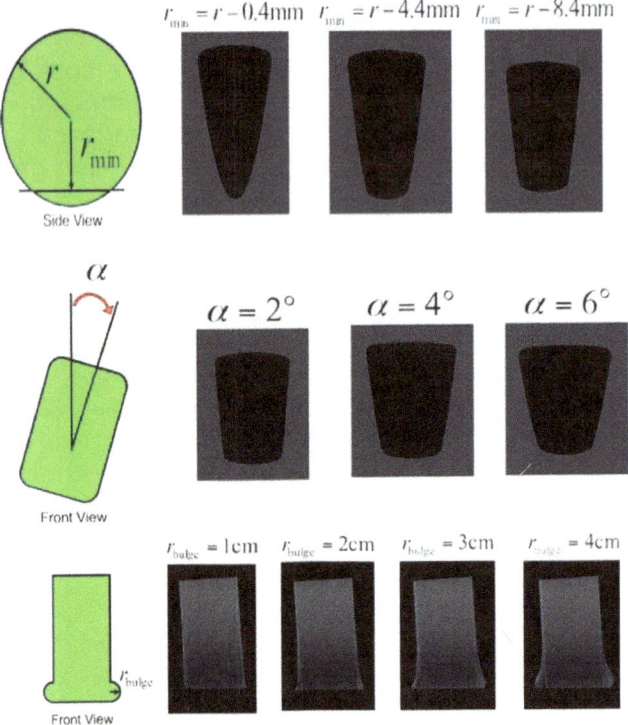

Impact of minimum radius (top) and camber angle (middle) on contact patch shape, and bulge radius (bottom) on front profile of the parametric tire

References

1] J. Anderson, Computational Fluid Dynamics, McGrawhill Inc, 1995.

2] J. Blazek, Computational Fluid Dynamics: Principles and Applications, Elsevier, 2005.

3] S. Patankar, Numerical Heat Transfer and Fluid Flow, CRC Press, 1980.

R. L. Meakin, Composite Overset

4] Structured Grids", Chapter 11, Handbook of Grid Generation, CRC Press, 1999.

5] O. C. Zienkiewicz, R. L. Taylor, O. C. Zienkiewicz and R. L. Taylor, The finite element method, London: McGraw-hill, 1977.

6] H. K. Versteeg and W. Malalasekera, An introduction to computational fluid dynamics: the finite volume method, Pearson Education, 2007.

7] R. D. Raucg, J. T. Batira and N. T. Y. Yang, "Spatial Adaption Procedures on Unstructured Meshes for Accurate Unsteady Aerodynamic Flow Computations.," AIAA, no. 91-1106, 1991.

8] D. G. Holmes and S. D. Conell, "Solution of the 2D Navier-Stokes Equations on Unstructured Adaptive Grids.," in AIAA 9th Computational Fluid Dynamics Conference, 1989.

9] D. C. Wilcox, Turbulence Modeling for CFD, D C W Industries, 2006.

10] F. R. Menter, AIAA Journal, vol. 8, no. 32, 1993.

11] G. Doig and T. J. Barber, "Considerations for Numerical Modeling of Inverted Wings in Ground Effect," AIAA Journal, vol. 49, no. 10, pp. 2330-2333, 2011.

12] J. Axerio-Ciles and I. Gianluca, "An aerodynamic investigation of an isolated rotating formula 1 wheel assembly.," ASME Journal of Fluids Engineering, vol. 12, no. 134, 2012.

H. Fiedler, "Coherent structures in 13] turbulent flows," *Prog. Aerospace Sci.*, vol. 25, pp. 231-269, 1988.

➔ Sample heat equation discretization – Excel sheet making:

The goal of this tutorial is to create an EXCEL spreadsheet that calculates the numerical solution to the following initial-boundary value problem for the one-dimensional heat equation:

$$\frac{\partial u}{\partial t} = \alpha^2 \frac{\partial^2 u}{\partial x^2} \qquad 0 < x < 1, \quad t > 0$$

BC: $u(0,t) = 0 \qquad u(1,t) = 0$

IC: $u(x,0) = f(x)$

The basic idea of the numerical approach to solving differential equations is to replace the derivatives in the heat equation by difference quotients and consider the relationships between u at (x,t) and its neighbours a distance Δx apart and at a time Δt later. In particular, in this tutorial the following expressions will be used:

- a *forward difference* in time:

$$\frac{u(x, t+\Delta t) - u(x,t)}{\Delta t} = \frac{\partial u(x,t)}{\partial t} + O(\Delta t)$$

- and a *central difference* in space:

$$\frac{u(x+\Delta x, t) - 2u(x,t) + u(x-\Delta x, t)}{\Delta x^2} = \frac{\partial^2 u(x,t)}{\partial x^2} + O(\Delta x^2)$$

By rewriting the heat equation in its discretized form using the expressions above and rearranging terms, one obtains:

$$u(x,t+\Delta t) = u(x,t) + \alpha^2\left(\frac{\Delta t}{\Delta x^2}\right)\left\{u(x+\Delta x,t) - 2u(x,t) + u(x-\Delta x,t)\right\}$$

Hence, given the values of u at three adjacent points $x-\Delta x$, x, and $x+\Delta x$ at a time t, one can calculate an approximated value of u at x at a later time $t+\Delta t$.

Using EXCEL spreadsheets allows you to perform these calculations repeatedly and effortlessly. The following are the necessary steps to set up a spreadsheet to calculate the solution to the initial-boundary value problem shown above.

Step 1: How to Tabulate the Spatial Interval

The first step is to tabulate the sample points in the **spatial interval**. Subdivide the interval [0, 1] into N+1 equally-spaced sample points a distance Δx apart.

In a blank EXCEL spreadsheet, enter the sample points along a row, as shown below. For this tutorial, let $\Delta x = 0.05$ and N = 19. Note that the value of Δx is shown in cell C2.

Each sample point is calculated as

$$x_n = x_{n-1} + \Delta x$$

for $n = 1, \ldots$ N and $x_0 = 0$.

For example, in the spreadsheet below the first sample point in the interval is $x_1 = x_0 + 0.05$. The value of x_1 is shown in cell E5: it is computed by addying the fixed space increment (cell C2) to the value contained in the cell to the immediate left of E5 (cell D5). The expression in the formula box in the image below shows the specific EXCEL formula used to compute the value in cell E5.

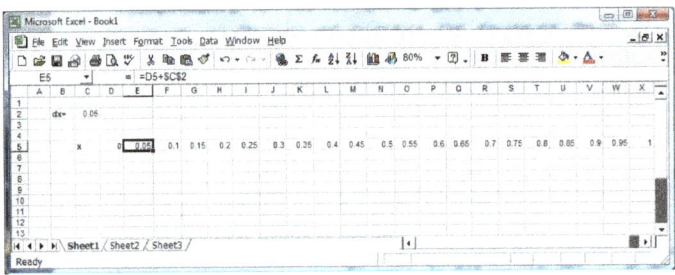

Step 2: How to Tabulate the Time Interval

The second step is to tabulate the **time interval**. Subdivide the interval [0,T] into M+1 equal time levels Δt long.

Enter the time levels in a column in the EXCEL spreadsheet, as shown below. For this tutorial, let $\Delta t = 0.004$ and M = 20. Note that the value of Δt is shown in cell F2.

Each time level is calculated as

$$t_k = t_{k-1} + \Delta t$$

for k = 1, ... M and $t_0 = 0$.

For example, in the spreadsheet below the second time level is $t_2 = t_1 + 0.004$.

The value of t_2 is shown in cell B9: it is computed by adding the fixed time increment (cell F2) to the value contained in the cell immediately above B9 (cell B8).

The expression in the formula box in the image below shows the specific EXCEL formula used to compute the value in cell B9.

Step 3: How to Include the Initial Condition

The next step is to assign the initial condition $u(x,0) = f(x)$. Note that the specific form of $f(x)$ depends on the characteristic of the problem under consideration. For this tutorial, choose

$f(x) = 2x$ for $0 < x < 0.5$ and $f(x) = 2-2x$ for $0.5 < x < 1$.

For clarity, in the spreadsheet below space-time points have been highlighted in blue and the corresponding values of the solution at $t = 0$ in green.

In other words, the values of the initial condition are shown in row 7, in cells D7 through X7. For example, the value of $u(0.1, 0)$, that is, the value of u at x_2 at t_0, is shown in cell F7.

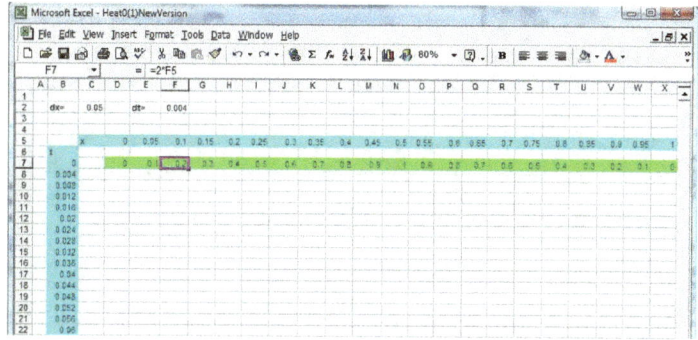

Step 4: How to Include the Boundary Conditions

The initial-boundary value problem discussed in this tutorial has two **boundary conditions**:

$u(0, t) = 0$ and $u(1, t) = 0$.

In the spreadsheet shown below, column D, from cells D7 through D27, contains the values corresponding to the first boundary condition $u(0, t) = 0$, that is, it shows the constant value of u at x_0. Column X, from cells X7 through X27, instead, contains the second boundary condition $u(1, t) = 0$, that is, it shows the constant value of u at x_N. Note that at t_0, $u(1, 0) = 0$ (cell X7) in agreement with the initial condition.

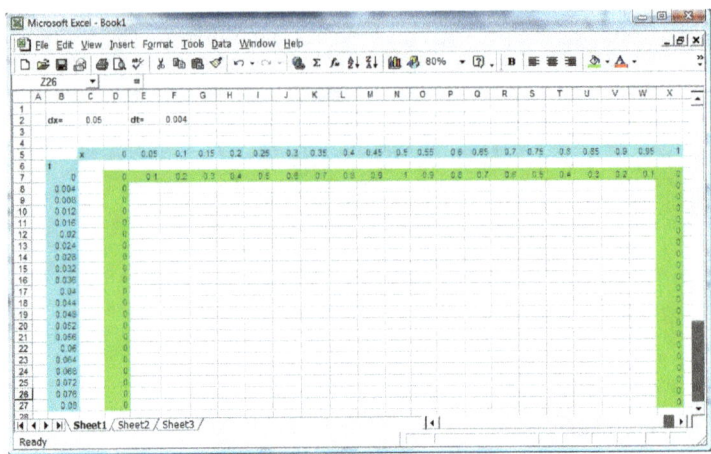

Step 5: How to Compute the Numerical Solution

To calculate the value of u at each space-time sample point (x_n, t_k), the following algorithm is used

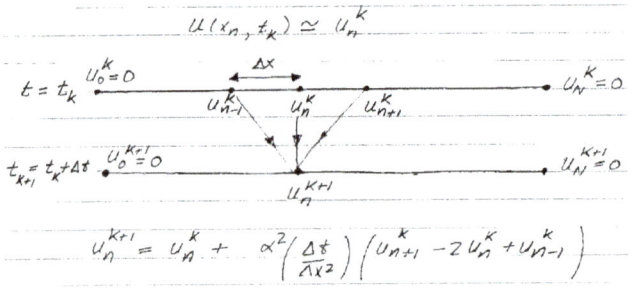

for $n = 1, 2, \ldots N$. Thus, u_n^{k+1}, the approximated value of u at point x_n at a time t_{k+1}, can be calculated given the values of u at three adjacent points, x_{n-1}, x_n, x_{n+1}, at a time t_k.

For example, the value of u at x_1 at time t_1 can be approximated as $u(x_1, t_1) \sim u_1^1$, and calculated using the expression above with $n = 1$ and $k = 0$,

$$u_1^1 = u_1^0 + K\left(u_2^0 - 2u_1^0 + u_0^0\right)$$

Where

$$K = \alpha^2 \left(\frac{\Delta t}{\Delta x^2}\right)$$

Hence, given the value of u at x_0, x_1, x_2 at time t_u, one can find an approximated value of u at x_1 one time step later. The spreadsheet below shows how to perform this calculation:

- the value of K is stored in cell O2
- the value of u_1^0 is stored in cell E7
- the value of u_2^0 is stored in cell F7
- the value of u_0^0 is stored in cell D7
- the value of u_1^1 is computed in cell E8 using the formula E7+O2*(F7-2*E7+D7), as displayed in the formula box

Note that cell O2 is called by absolute referencing (O2 instead of O2). This is because the formula used in cell E8 will be copied and pasted in other cells, as explained later on. It is worth noticing that the calculation performed in cell E8 uses the numerical values stored in three cells in the row just above it, specifically in cell E7, the cell immediately *above* E8, and cells D7 and F7, the cells to the *left* and *right* of E7, respectively. This way of storing information in rows and columns allows you to organize your computation in a structured manner. As you fill in the spreadsheet to calculate *u* over the entire domain, information will always be retrieved from cells one row above the row you are working on, in the same column and in the columns to the immediate left and right.

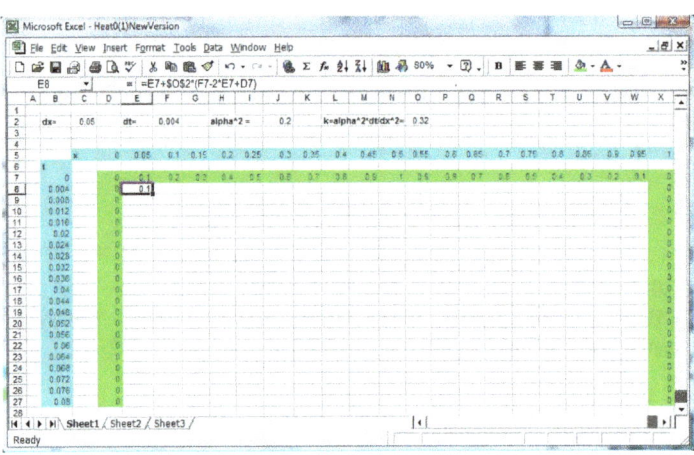

One could perform a similar calculation to find u_2^1, the (approximated) value of u at a sample point a distance Δx to the right of x_1 and at the same time level as u_1^1. The result of the computation would be stored in the same row (or time level) as u_1^1 (row 8), but one column to the right (column F), to be consistent with the way sample space-time points are tabulated in the spreadsheet. Furthermore, the computation of u_2^1 would involve the values of u at x_1, x_2, and x_3 at time t_0, which are stored in cells E7, F7, and G7. In other words, the computation of u_2^1 would involve information in three cells located to the immediate right with respect to the cells used to compute u_1^1. Therefore, instead of typing a new formula for u_2^1 in cell F8, one could simply **Copy** and **Paste** the formula from cell E8 into cell F8, EXCEL will then automatically update the formula. In fact, when a formula is copied from one cell to another, EXCEL does not create an exact copy of the formula, instead it changes cell addresses relative to the row and column they are moved to. When the formula from cell E8 is pasted one cell to the right (into F8), all column labels in the formula, with the exception of cell O2, will be shifted to the right, and the correct values of u will be accessed. The cell address for O2 will not change because the column and row labels are preceded by the "$" sign (this is called "absolute referencing" of cells), allowing for the constant value of K to be read always from cell O2.

The same method can be used to calculate values of u at later times. Instead of typing a new formula in subsequent rows, simply copy and paste the formula from the cell corresponding to the previous time level, i.e. the cell in the row above.

Keep in mind that in order to compute the time evolution of u at a particular point in the space interval, its earlier value at that point and at adjacent points must be known. Therefore, before calculating how u changes with time at a given point, it is convenient to find earlier values of u at all points in the spatial interval. In other words, when using the spreadsheet to calculate u, first fill in all cells in one row, and then move to the row below, so that the necessary values of u are always available. The **Edit|Fill|Fill Right** and **Edit|Fill|Fill Down** features will allow you to fill in all cells and calculate the solution quickly.

Step 6: How to Plot the Solution

The time evolution of u can be easily observed if one plots the solution computed in the spreadsheet at different time levels, and compare it with the initial condition. Here is a sequence of steps to make a plot of the solution at time $t = 0$ and at time $t = 0.08$ (corresponding to M = 20).

- Select with the mouse all cells in row 5, from D5 to X5, by left-clicking and scrolling along the row.
- Press the CTRL key and hold it.
- Select with the mouse all cells in row 7, from D7 to X7, by left-clicking and scrolling along the row.
- start the Chart Wizard by clicking on the icon on the toolbar.
- Select the "XY (Scatter)" chart type and a chart sub-type of your choice.
- Select "Series in Rows" and press "Next".
- Enter a chart title and axis labels, if desired.

- Place chart "as object in Sheet 1" if you want your plot to be displayed in the same worksheet as your data.
- Click "Finish" to creat the plot of $u(x)$ at $t = 0$.
- Right-click with the mouse on any data point on the chart and select "Source Data" from the menu.
- Type a name for the data series in the chart, for example "u(x) at t = 0". This is the description that will appear on the chart legend for this data series.
- Click "Add" series to add a new curve to the same plot, for example to plot $u(x)$ at $t = 0.08$.
- type a name for the new curve, for example "u(x) at t = 0.08".
- Left-click in the dialog box for X Values and select all cells in row 5 from D5 through X5.
- Delete any text present in the dialog box for Y Values and select with the mouse (righ-click and hold) all cells in row 27 from D27 through X27.
- Click OK to add a second curve to your plot.

Your plot may look like the image below. By increasing M and implementing the computation on more time level (i.e., by filling more rows in the spreadsheet), you can investigate the evolution of u at even later time.

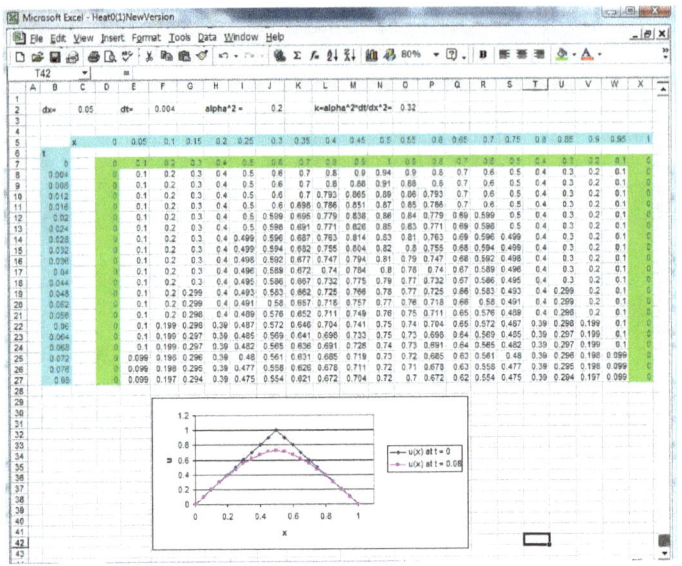

Step 7: How to Implement Derivative Boundary Conditions

Suppose the boundary conditions to the initial-boundary value problem discussed at the beginning of this tutorial are changed to

$$u(0,t) = 0 \quad , \qquad \frac{\partial u}{\partial x}(1,t) = 0$$

Unlike the previous case, in which the boundary conditions provided the value of the solution on both boundaries of the domain, now the value of the solution is known only at $x = 0$. Thus, $u(1, t)$ must be calculated using the same difference approximation for the heat equation used in Step 5, but now $n = N$, that is,

- Place chart "as object in Sheet 1" if you want your plot to be displayed in the same worksheet as your data.
- Click "Finish" to creat the plot of $u(x)$ at $t = 0$.
- Right-click with the mouse on any data point on the chart and select "Source Data" from the menu.
- Type a name for the data series in the chart, for example "u(x) at t = 0". This is the description that will appear on the chart legend for this data series.
- Click "Add" series to add a new curve to the same plot, for example to plot $u(x)$ at $t = 0.08$.
- type a name for the new curve, for example "u(x) at t = 0.08".
- Left-click in the dialog box for X Values and select all cells in row 5 from D5 through X5.
- Delete any text present in the dialog box for Y Values and select with the mouse (righ-click and hold) all cells in row 27 from D27 through X27.
- Click OK to add a second curve to your plot.

Your plot may look like the image below. By increasing M and implementing the computation on more time level (i.e., by filling more rows in the spreadsheet), you can investigate the evolution of u at even later time.

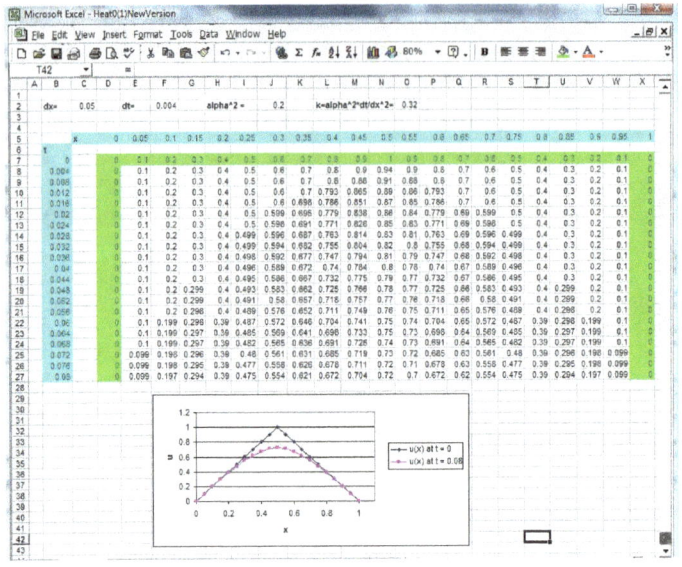

Step 7: How to Implement Derivative Boundary Conditions

Suppose the boundary conditions to the initial-boundary value problem discussed at the beginning of this tutorial are changed to

$$u(0,t) = 0 \, , \qquad \frac{\partial u}{\partial x}(1,t) = 0$$

Unlike the previous case, in which the boundary conditions provided the value of the solution on both boundaries of the domain, now the value of the solution is known only at $x = 0$. Thus, $u(1, t)$ must be calculated using the same difference approximation for the heat equation used in Step 5, but now $n = N$, that is,

$$u_N^{k+1} = u_N^k + \alpha^2\left(\frac{\Delta t}{\Delta x^2}\right)\left(u_{N+1}^k - 2u_N^k + u_{N-1}^k\right)$$

By doing so, one finds that the value of u at x_N depends on the value of u at x_{N+1}. Recall that $x_{N+1} = x_N + \Delta x$. Since $x_N = 1$, then, x_{N+1} is outside the domain and must be defined. In practical terms, this means adding a column in the spreadsheet used to calculate the solution. Alternatively, one could seek a new expression for u_N^k that does not depend on the value of u at x_{N+1}, avoiding therefore the additional column in the spreadsheet. In both cases, it will be essential to use the information provided by the derivative boundary condition, as described below.

Consider a central difference approximation of the derivative boundary condition,

$$\frac{u(x_N + \Delta x, t) - u(x_N - \Delta x, t)}{\Delta x} = 0$$

It follows that

$$u(x_N + \Delta x, t) = u(x_N - \Delta x, t)$$

Since $x_{N+1} = x_N + \Delta x$ and $x_{N-1} = x_N - \Delta x$, it follows that $u_{N+1}^k = u_{N-1}^k$. This information can be used to calculate u_N^{k+1} in two ways:

- either eliminate u_{N+1}^k from the expression for u_N^{k+1} using the fact that $u_{N+1}^k = u_{N-1}^k$, and then in column X of the spreadsheet implement the new expression

$$u_N^{k+1} = u_N^k + 2\alpha^2 \left(\frac{\Delta t}{\Delta x^2}\right)\left(u_{N-1}^k - u_N^k\right)$$

- or add an extra column in the spreadsheet, e. g. column Y, into which the values from column W, which corresponds to x_{N-1}, are copied, and then in column X implement the usual expression

$$u_N^{k+1} = u_N^k + \alpha^2 \left(\frac{\Delta t}{\Delta x^2}\right)\left(u_{N+1}^k - 2u_N^k + u_{N-1}^k\right)$$

www.ingramcontent.com/pod-product-compliance
Lightning Source LLC
Chambersburg PA
CBHW072148230526
45467CB00041B/954